DICTIONARY OF CONSTRUCTION ENGINEERING AND MANAGEMENT

Steven Smith, Ph.D.

Wisdom Publishers

To the pioneers and innovators who have shaped the landscape of construction engineering and management, whose dedication and ingenuity have transformed our built environment. May this dictionary serve as a testament to their unwavering commitment to excellence and inspire future generations to continue pushing the boundaries of this dynamic field.

Building is the greatest art, for which all other arts are preliminary and ancillary.

JOHN RUSKIN

CONTENTS

INTRODUCTION

The construction industry plays a pivotal role in shaping our physical world, transforming raw materials into the infrastructure and buildings that support modern society. From the towering skyscrapers that define our cities to the intricate networks of roads and bridges that connect communities, construction projects shape the landscape we inhabit and the lives we lead.

The Dictionary of Construction Engineering and Management is a comprehensive resource designed to provide a clear, concise, and accessible understanding of the terminology, concepts, and practices that underpin this dynamic field. Compiled with the needs of professionals, students, and anyone interested in construction in mind, this dictionary serves as a gateway to the vast and ever-evolving world of construction engineering and management.

Within these pages, you will find a comprehensive collection of terms encompassing the breadth of construction engineering and management, from the fundamental principles of materials science and structural analysis to the intricacies of project planning, cost control, and sustainable construction practices. Each entry is carefully crafted to provide clear and concise definitions, supplemented by synonyms, antonyms, and related terms to enhance understanding and expand your vocabulary.

To further enrich your learning experience, the dictionary incorporates cross-references that guide you seamlessly between related concepts, allowing you to explore the interconnectedness of construction knowledge. Additionally, illustrations, diagrams, and other visuals are integrated throughout the dictionary to elucidate complex concepts and demonstrate the application of terms in real-world scenarios.

The Dictionary of Construction Engineering and Management is more than just a collection of definitions; it is an essential tool for anyone seeking to navigate the complexities of construction engineering and management. Thank you for buying this dictionary. I believe that it will serve as your guide.

PART I: FUNDAMENTALS OF CONSTRUCTION ENGINEERING AND MANAGEMENT

CHAPTER 1: CONSTRUCTION PROJECT MANAGEMENT PRINCIPLES AND PRACTICES

Budgeting | The process of estimating, allocating, and managing financial resources for a construction project. Budgeting involves identifying the costs associated with labor, materials, equipment, and overhead, ensuring that expenditures align with the project's financial plan.

Building Information Modeling (BIM) | A sophisticated digital tool that facilitates collaborative and efficient planning, design, construction, and management of projects. BIM creates a 3D model that integrates various project data, enhancing communication among stakeholders and minimizing errors during the construction phase.

Change Order | A formal request to modify the original scope of work in a construction project. Change orders can involve changes in design, materials, or other project elements and are essential for addressing unforeseen issues or accommodating client-requested alterations.

Communication Plan | A strategic document outlining how project information will be shared among team members and stakeholders. A communication plan ensures timely and effective dissemination of critical project updates, fostering collaboration and preventing misunderstandings.

Construction Manager at Risk (CMAR) | A project delivery method where the construction manager is involved early in the design phase and assumes certain project risks. This collaborative approach allows for enhanced communication, cost control, and efficient problem-solving throughout the project.

Construction Phase | The stage of a project where physical construction work takes place on the site. During this phase, project plans are translated into tangible structures, and coordination among contractors, subcontractors, and other stakeholders is crucial for successful execution.

Cost Estimation | The process of predicting the expenses associated with a construction project, considering materials, labor, equipment, permits, and other relevant costs. Accurate cost estimation is essential for budgeting and project planning.

Cost Overrun | Occurs when the actual costs of a project exceed the budgeted or estimated costs. Cost overruns can result from unexpected challenges, changes in project scope, or inadequate cost estimation.

Cost Performance Index (CPI) | A measure of cost efficiency on a project, calculated as the ratio of earned value to actual cost. A CPI value greater than 1 indicates favorable cost performance, while a value less than 1 signals cost overruns.

Design-Bid-Build (DBB) | A traditional project delivery method where the design phase is completed before bidding and construction. This sequential approach allows for a competitive bidding process but may result in longer project timelines.

Design-Build (DB) | A project delivery method where the design and construction phases are contracted to a single entity. This integrated approach streamlines communication, accelerates project delivery, and encourages collaboration between designers and builders.

Design Construction Management (DCM) | An integrated approach combining design and construction management services. DCM emphasizes collaboration between design and construction professionals from project inception, leading to more efficient project delivery.

Design Development (DD) | The phase in project development where the design is refined and detailed before moving into the construction phase. DD involves further specifying materials, finishes, and technical details to ensure alignment with project objectives.

Earned Value Management (EVM) | A project management technique that assesses a project's performance against its baseline schedule and budget. EVM integrates project scope, schedule, and cost, providing insights into project health and forecasting.

Engineering Procurement Construction (EPC) | A project delivery method where a single entity is responsible for engineering, procurement, and construction. EPC contracts simplify project management by centralizing responsibility for the entire project life cycle.

Environmental Impact Assessment (EIA) | An evaluation of the potential environmental effects of a project before it begins. EIA involves identifying, predicting, and assessing the impact of project activities on the environment.

Execution Phase | The stage of a project where planned work is carried out, and construction activities commence. Execution involves the physical realization of the project plan and requires effective coordination and monitoring.

Gantt Chart | A visual representation of a project schedule, displaying tasks, durations, and dependencies. Gantt charts help project managers and stakeholders understand the timeline and sequence of project activities.

Green Building | Construction practices aimed at reducing a project's environmental impact. Green building incorporates sustainable materials, energy-efficient design, and eco-friendly construction techniques.

Guaranteed Maximum Price (GMP) | The highest price a contractor agrees to charge for a project, often used in design-build contracts. GMP provides cost certainty for the client, with any cost overruns borne by the contractor.

Integrated Project Delivery (IPD) | A collaborative project delivery approach where all stakeholders work together from project inception. IPD promotes shared responsibility, risk, and reward among project participants.

Key Performance Indicators (KPIs) | Metrics used to evaluate the success of a project in meeting its objectives. KPIs provide quantifiable measures of project performance and can include factors such as cost efficiency, schedule adherence, and quality.

Lean Construction | A production management approach focused on maximizing efficiency and minimizing waste in construction projects. Lean construction aims to eliminate non-value-added activities and improve overall project performance.

Life Cycle Cost Analysis (LCCA) | An assessment of all costs associated with a project over its entire life cycle, including construction, operation, and maintenance. LCCA informs decision-making by considering the total cost of ownership.

Milestone | A significant event or point in time during a project. Milestones mark achievements, completion of phases, or other critical points, helping track project progress.

Monitoring and Controlling Phase | The part of project management that involves tracking, reviewing, and regulating the progress and performance of the project. Monitoring and controlling ensure that the project stays on course and objectives are met.

Performance Bond | A guarantee that a contractor will fulfill their contractual obligations, often backed by a financial institution. Performance bonds provide assurance to project owners that the contractor will complete the project as specified.

Procurement | The process of acquiring goods, services, or works from external sources for a construction project. Procurement involves selecting suppliers, negotiating contracts, and ensuring the timely delivery of materials and services.

Project Charter | A document that formally authorizes a project and provides the project manager with the authority to use organizational resources for project activities. The project charter outlines project goals, objectives, and stakeholders.

Project Closeout | The final phase of a project where all aspects are completed, and the project is formally closed. Closeout includes final inspections, documentation, and transitioning deliverables to the client.

Project Life Cycle | The stages a project goes through from initiation to closure. Project life cycle phases typically include initiation, planning, execution, monitoring and controlling, and closure.

Project Manager | The individual responsible for planning, executing, and closing a construction project. The project manager oversees all aspects of the project, ensuring it meets its goals within scope, time, and budget.

Project Scope | The detailed description of the project and its deliverables. Project scope defines what will and will not be included in the project and serves as a basis for planning and execution.

Quality Assurance (QA) | A set of planned and systematic activities to ensure that a project meets specified requirements. QA focuses on preventing defects and errors in project deliverables.

Quality Control (QC) | The process of inspecting and testing project deliverables to ensure they meet quality standards. QC identifies and corrects defects, ensuring that project outputs meet the defined quality criteria.

Request for Proposal (RFP) | A document used to solicit proposals from potential contractors or suppliers. An RFP outlines project requirements, allowing contractors to submit detailed proposals for consideration.

Risk Management | The process of identifying, assessing, and controlling project risks to minimize potential negative impacts. Risk management involves proactive planning to address uncertainties that may affect the project.

Schedule Performance Index (SPI) | A measure of schedule efficiency on a project, showing the ratio of earned value to planned value. SPI values greater than 1 indicate favorable schedule performance.

Scope Creep | The gradual expansion or modification of a project's scope without proper approval. Scope creep can lead to schedule delays and increased costs.

Stakeholder | Any individual or group that can affect or is affected by a construction project. Stakeholders may include clients, project team members, regulators, and the community.

Subcontractor | A company or individual hired by a general contractor to perform specific tasks on a construction project. Subcontractors specialize in particular trades or services.

Sustainability | Practices that aim to reduce a project's impact on the environment and promote long-term viability. Sustainable construction includes energy-efficient design, use of renewable materials, and environmentally conscious building practices.

Work Breakdown Structure (WBS) | A hierarchical decomposition of the total scope of work to be carried out by the project team. The WBS breaks down the project into manageable components, facilitating planning and resource allocation.

CHAPTER 2: CONSTRUCTION PLANNING, SCHEDULING, AND COST CONTROL

Activity | A specific task or job that needs to be done during a construction project. For example, laying the foundation or installing windows. Example: Constructing a building involves various activities, such as excavation, foundation laying, and erecting structural elements.
Cross-reference: See also Work Package for a small, manageable part of a construction project that can be assigned to a specific team or individual.

Alternative Dispute Resolution (ADR) | A way to resolve disagreements or conflicts in construction projects without going to court. It includes methods like talking through issues or using a mediator. Example: In case of disputes during a construction project, parties may opt for ADR methods like mediation or arbitration instead of pursuing legal action. Cross-reference: Refer to Claim for a formal request for compensation or adjustment due to unexpected issues in a construction project.

As-Built Schedule | A record that shows how the construction project actually progressed compared to the original plan. It helps understand what was done and when. Example: The as-built schedule provides a chronological record of completed tasks, offering insights into project evolution. Cross-reference: See also Baseline Schedule for the initial plan outlining how the construction project should happen.

Bar Chart | A simple visual tool that uses bars to show the timing and duration of different tasks in a construction project. It's an easy way to see the project timeline. Example: A bar chart visually represents tasks like excavation, framing, and finishing, displaying their duration and overlap. Cross-reference: Refer to Gantt Chart for another visual tool used in project scheduling.

Baseline Schedule | The initial plan that outlines how the construction project should happen before any work begins. It's like a roadmap for the project. Example: The baseline schedule serves as a guide, detailing the sequence and timing of tasks before the start of construction. Cross-reference: See also As-Built Schedule for a record that shows how the construction project actually progressed.

Benchmarking | Comparing how well a construction project is doing against industry standards or best practices. It helps to see if the project is on track or if there are areas to improve. Example: Benchmarking involves evaluating construction performance against industry benchmarks, identifying areas for improvement. Cross-reference: Refer to Key Performance Indicators (KPIs) for metrics used to measure the success of a construction project.

Bid | An official offer from a construction company to do a specific project for a certain amount of money. Bids are like proposals for the work. Example: The bid submitted by a construction company outlines the cost, timeline, and details of the proposed project. Cross-reference: See also Procurement for the process of acquiring materials, services, or labor for a construction project.

Budget | The financial plan for a construction project that estimates how much money is needed and where it will be spent. It's like a financial roadmap for the project. Example: Creating a budget involves estimating costs for materials, labor, permits, and other expenses essential for the construction project. Cross-reference: Refer to Cash Flow for the movement of money in and out of a construction project.

Cash Flow | The movement of money in and out of a construction project. Managing cash flow is important to ensure there's enough money to pay for everything. Example: Proper cash flow management ensures that funds are available when needed, covering expenses like materials, labor, and permits. Cross-reference: See also Budget for the financial plan that estimates how much money is needed for a construction project.

Change Management | Dealing with changes to the original plan of a construction project. It involves assessing how changes will impact time, cost, and resources. Example: Change management evaluates the effects of alterations to the construction plan, considering adjustments in time, budget, and resources. Cross-reference: Refer to Change Order for a formal request to change something in the original plan of a construction project.

Change Order | A formal request to change something in the original plan of a construction project. For example, if the client wants to add a pool, a change order is initiated. Example: A change order may involve modifying the original design to accommodate additional rooms or altering the project timeline. Cross-reference: See also Claim for a formal request for compensation or adjustment due to unexpected issues in a construction project.

Claim | A formal request for compensation or adjustment due to unexpected issues or changes in a construction project. It's like saying, "We need more time or money because of this problem." Example: If unexpected challenges arise, a construction team may file a claim to seek compensation or adjustments to the project plan. Cross-reference: Refer to Alternative Dispute Resolution (ADR) for methods to resolve disagreements without going to court.

Communication Plan | A strategy outlining how information will be shared among everyone involved in the construction project. It ensures that everyone stays informed. Example: A communication plan establishes protocols for sharing project updates, milestones, and critical information among project stakeholders. Cross-reference: See also Stakeholder for anyone who is affected by or can affect a construction project.

Constructability | Assessing how easy or difficult it will be to build a project based on the design. It helps identify potential construction challenges. Example: Constructability analysis evaluates the ease of implementing the design, highlighting potential construction obstacles. Cross-reference: Refer to Risk Management for identifying, assessing, and mitigating potential risks in a construction project.

Construction Phase Plan | A detailed plan for the construction stage of a project. It includes timelines, resources, and safety measures. Example: The construction phase plan outlines specific activities, timelines, safety protocols, and resource requirements for the construction stage. Cross-reference: See also Master Schedule for the overall plan that outlines the sequence and timing of all tasks in a construction project.

Contingency | Extra money set aside in the budget for unforeseen issues or changes during construction. It's like a safety net for unexpected costs. Example: Contingency funds provide a financial buffer to address unforeseen challenges, ensuring the construction project stays on track. Cross-reference: Refer to Risk Mitigation for taking actions to minimize the impact of potential risks in a construction project.

Cost Control | Managing and keeping track of expenses to make sure the construction project stays within budget. It involves monitoring costs closely. Example: Cost control mechanisms include regular monitoring of expenses to ensure adherence to the planned budget throughout the construction project. Cross-reference: See also Cost Performance Index (CPI) for a measure of how well a construction project is staying within budget.

Cost Estimate | An educated guess of how much money will be needed for a construction project. It helps in creating the budget. Example: A cost estimate involves assessing the anticipated expenses for labor, materials, permits, and other factors essential for the construction project. Cross-reference: Refer to Estimation for the process of predicting how much time and money will be needed for a construction project.

Cost Overrun | When the actual costs of a construction project end up being more than what was budgeted. It's like going over the planned budget. Example: Cost overruns occur when unforeseen expenses cause the construction project's total cost to exceed the initially budgeted amount. Cross-reference: See also Cost Performance Index (CPI) for a measure of how well a construction project is staying within budget.

Cost Performance Index (CPI) | A measure of how well a construction project is staying within budget. It compares the planned cost to the actual cost. Example: A CPI value greater than 1 indicates that the construction project is under budget, while a value less than 1 signals a cost overrun. Cross-reference: Refer to Cost Control for managing and keeping track of expenses to ensure adherence to the construction project budget.

Critical Path | The sequence of tasks in a construction project that determines the shortest time for project completion. It helps in scheduling and managing time effectively. Example:

Identifying the critical path is crucial for scheduling, as it pinpoints tasks that, if delayed, would extend the overall construction project timeline. Cross-reference: See also Network Analysis for a method of planning and scheduling construction tasks based on their relationships and dependencies.

Deliverable | A specific item or task that needs to be completed and delivered in a construction project. It's like a project milestone. Example: In construction, deliverables include completed phases or specific tasks, such as the foundation being poured or the roof being installed. Cross-reference: Refer to Milestone for a significant point or event in a construction project that marks a key achievement or completion of a major task.

Design Development (DD) | The phase where the initial design is refined before starting construction. It involves finalizing details. Example: Design development ensures that all design details, from architectural elements to engineering specifications, are thoroughly finalized. Cross-reference: See also Estimation for the process of predicting how much time and money will be needed for a construction project.

Earned Value | A measure of the value of work completed in a construction project. It helps assess project progress. Example: Earned value is calculated by assessing the completed work's value against the planned project schedule and budget. Cross-reference: Refer to Earned Value Management (EVM) for a technique that combines cost, schedule, and scope to evaluate a construction project's performance.

Earned Value Management (EVM) | A technique that combines cost, schedule, and scope to evaluate a construction project's performance. It provides insights into how well the project is doing. Example: EVM involves assessing the value of work completed, comparing it to the planned value, and providing insights into project performance. Cross-reference: See also

Critical Path for the sequence of tasks determining the shortest time for project completion.

Estimation | The process of predicting how much time and money will be needed for a construction project. It's an educated guess based on available information. Example: Estimation involves using historical data, industry standards, and expert judgment to predict costs and timelines for a construction project. Cross-reference: Refer to Cost Estimate for an educated guess of how much money will be needed for a construction project.

Fast-Tracking | Completing tasks in a construction project as quickly as possible to shorten the overall project duration. It involves overlapping activities. Example: Fast-tracking may involve starting the construction of certain elements before the entire design is finalized to expedite project completion. Cross-reference: See also Schedule Compression for shortening the overall duration of a construction project without changing the project scope.

Float | The amount of time that a task or project can be delayed without affecting the overall project timeline. It provides flexibility in scheduling. Example: Float allows for the delay of non-critical tasks without impacting the construction project's final completion date. Cross-reference: Refer to Total Float for the amount of time that a task or project can be delayed without affecting the overall project timeline.

Forensic Schedule Analysis | Investigating and analyzing the construction schedule after a project is completed. It helps understand what went well or could be improved. Example: Forensic schedule analysis involves a thorough examination of the construction schedule to identify strengths, weaknesses, and areas for improvement. Cross-reference: See also Network Analysis for a method of planning and scheduling construction tasks based on their relationships and dependencies.

Gantt Chart | A visual tool that shows the timeline of tasks in a construction project using horizontal bars. It provides a clear view of the project schedule. Example: A Gantt chart visually represents the start and end dates of various construction tasks, aiding in project schedule visualization. Cross-reference: Refer to Bar Chart for another visual tool used to show the timing and duration of different tasks in a construction project.

Human Resources | The people involved in a construction project, including workers, managers, and support staff. Managing human resources ensures the right people are in the right roles. Example: Human resources management in construction involves assigning skilled workers to tasks, ensuring efficient project execution. Cross-reference: See also Team Building for fostering cooperation and collaboration among individuals working on a construction project.

Independent Estimate | A separate and unbiased calculation of the cost of a construction project. It helps verify the accuracy of cost estimates. Example: An independent estimate is conducted by a third party to ensure objectivity and accuracy in assessing the cost of a construction project. Cross-reference: Refer to Estimation for the process of predicting how much time and money will be needed for a construction project.

Integrated Master Plan (IMP) | A comprehensive plan that combines all project plans into a single, cohesive document. It ensures that all aspects of the project are aligned. Example: The integrated master plan brings together schedules, resources, and milestones, providing a holistic view of the construction project. Cross-reference: See also Master Schedule for the overall plan that outlines the sequence and timing of all tasks in a construction project.

Just-In-Time (JIT) | A strategy to minimize inventory and avoid unnecessary costs by delivering materials exactly when they are needed for construction. It's like ordering materials only when they are required. Example: Just-In-Time delivery ensures that construction materials are delivered precisely when needed, minimizing storage costs and waste.Cross-reference: Refer to Procurement for the process of acquiring materials, services, or labor for a construction project.

Key Performance Indicators (KPIs) | Metrics used to measure the success of a construction project. KPIs provide a way to track progress and performance. Example: KPIs in construction may include metrics like on-time project completion, adherence to budget, and client satisfaction. Cross-reference: See also Benchmarking for comparing construction project performance against industry standards or best practices.

Lag Time | The delay between the finish of one task and the start of the next in a construction project. It helps in sequencing activities. Example: Lag time is incorporated in project scheduling to ensure that the completion of one task aligns with the start of the next. Cross-reference: Refer to Lead Time for the time required to prepare for and start a construction task once it's initiated.

Lead Time | The time required to prepare for and start a construction task once it's initiated. It helps in planning and scheduling. Example: The lead time for ordering specialized construction materials is crucial to ensure they arrive on-site when needed. Cross-reference: Refer to Lag Time for the delay between the finish of one task and the start of the next in a construction project.

Lean Construction | An approach that focuses on minimizing waste and maximizing efficiency in construction projects. It aims to optimize resources. Example: Lean construction principles involve streamlining processes to eliminate unnecessary steps, reducing costs and time. Cross-reference: See also Value Engineering for optimizing the value of a construction project.

Life-Cycle Cost | The total cost of a construction project, including initial costs, operating costs, and maintenance costs over its entire life. It helps in evaluating long-term expenses. Example: Considering life-cycle costs involves evaluating not only initial construction expenses but also long-term maintenance and operational costs. Cross-reference: Refer to Life-Cycle Cost Analysis (LCCA) for evaluating all costs associated with a construction project over its entire life.

Life-Cycle Cost Analysis (LCCA) | Evaluating all costs associated with a construction project over its entire life. It guides decision-making by considering the project's total cost. Example: Life-cycle cost analysis considers factors such as maintenance, energy consumption, and replacement costs to make informed decisions. Cross-reference: See also Life-Cycle Cost for the total cost of a construction project over its entire life.

Logical Relationship | The connection between different tasks in a construction project that defines their sequence. It helps in creating a logical project schedule. Example: Logical relationships determine the order of tasks, ensuring that one activity logically follows another in the construction process. Cross-reference: Refer to Network Diagram for a visual representation of tasks and their relationships in a construction project.

Master Schedule | The overall plan that outlines the sequence and timing of all tasks in a construction project. It serves as a guide for project managers. Example: The master schedule provides an overview of the entire construction project, including milestones, deliverables, and deadlines. Cross-reference: See also Baseline Schedule for the initial plan outlining how the construction project should happen.

Milestone | A significant point or event in a construction project that marks a key achievement or completion of a major task. It's like a project checkpoint. Example: Pouring the foundation can be a milestone in a construction project, signifying a critical phase completion. Cross-reference: Refer to Activity for a specific task or job in a construction project.

Network Analysis | A method of planning and scheduling construction tasks based on their relationships and dependencies. It helps in creating an efficient project schedule. Example: Network analysis identifies critical paths and dependencies, optimizing task sequencing for efficient project completion. Cross-reference: See also Critical Path for the sequence of tasks determining the shortest time for project completion.

Network Diagram | A visual representation of tasks and their relationships in a construction project. It provides a clear picture of the project's flow. Example: A network diagram visually displays task relationships, helping project managers understand the project's overall structure. Cross-reference: Refer to Gantt Chart for another visual tool used in project scheduling.

Owner-Controlled Insurance Program (OCIP) | An insurance program controlled by the project owner that provides coverage for construction-related risks. It simplifies insurance management. Example: OCIP consolidates insurance coverage under the project owner, reducing complexities in insurance management for various contractors. Cross-reference: See also Insurance for overall risk mitigation strategies.

PERT (Program Evaluation and Review Technique) | A technique for estimating the duration of tasks in a construction project, considering best-case, worst-case, and most likely scenarios. It helps create a more realistic schedule. Example: PERT analysis considers various scenarios for each task, providing a more accurate estimate of project duration. Cross-reference: Refer to Critical Path for the sequence of tasks determining the shortest time for project completion.

Performance Bond | A guarantee from a contractor that they will complete a construction project as specified in the contract. It provides assurance to the project owner. Example: A performance bond assures the project owner that the contractor will fulfill contractual obligations, providing financial security. Cross-reference: See also Contract for legal agreements outlining project specifications and terms.

Procurement | The process of acquiring materials, services, or labor for a construction project. It involves selecting suppliers and negotiating contracts. Example: Procurement includes sourcing construction materials, hiring subcontractors, and negotiating contracts to meet project needs. Cross-reference: Refer to Bid for an official offer from a construction company to do a specific project.

Project Management Plan | A detailed document that outlines how a construction project will be executed, monitored, and controlled. It serves as a guide for project managers. Example: The project management plan details project scope, schedules, budgets, and risk management strategies, providing a comprehensive guide. Cross-reference: See also Construction Phase Plan for a detailed plan specific to the construction stage of a project.

Quality Management | Ensuring that the construction project meets the specified standards and requirements. It involves processes to maintain quality throughout the project. Example: Quality management involves inspections, testing, and adherence to industry standards to deliver a high-quality construction project. Cross-reference: Refer to Constructability for assessing how easy or difficult it will be to build a project based on the design.

Request for Information (RFI) | A formal request for clarification or additional information about a construction project. It helps in addressing uncertainties. Example: An

RFI may seek clarification on design details, materials, or construction methods to ensure accurate project execution. Cross-reference: See also Communication Plan for strategies outlining how information will be shared among everyone involved in the construction project.

Resource Allocation | Assigning the right people, equipment, and materials to tasks in a construction project. It ensures that resources are used efficiently. Example: Resource allocation involves assigning skilled workers, appropriate machinery, and necessary materials to specific project tasks. Cross-reference: Refer to Resource Leveling for balancing the workload and resources in a construction project.

Resource Leveling | Balancing the workload and resources in a construction project to avoid overloading or underutilization. It helps in maintaining a steady pace of work. Example: Resource leveling ensures an even distribution of tasks to prevent bottlenecks or resource shortages during construction. Cross-reference: See also Resource Allocation for assigning the right people, equipment, and materials to tasks.

Risk Management | Identifying, assessing, and mitigating potential risks in a construction project. It involves planning for uncertainties that could impact the project. Example: Risk management assesses potential issues such as weather delays, supply chain disruptions, and design changes, developing strategies to mitigate their impact. Cross-reference: Refer to Risk Mitigation for taking actions to minimize the impact of potential risks in a construction project.

Risk Mitigation | Taking actions to minimize the impact of potential risks in a construction project. It involves proactive planning to reduce the likelihood of negative events. Example: Risk mitigation strategies may include developing alternative plans for critical tasks or securing backup suppliers to address potential issues. Cross-reference: See also Risk Management for identifying, assessing, and mitigating potential risks.

Schedule | A plan that outlines when different tasks in a construction project will be performed. It provides a timeline for project activities. Example: The construction schedule details when excavation, foundation pouring, framing, and other tasks will occur. Cross-reference: Refer to Master Schedule for the overall plan that outlines the sequence and timing of all tasks in a construction project.

Schedule Compression | Shortening the overall duration of a construction project without changing the project scope. It involves finding ways to complete tasks more quickly. Example: Schedule compression may involve overlapping tasks or using faster construction methods to meet project deadlines. Cross-reference: See also Fast-Tracking for completing tasks in a construction project as quickly as possible.

Schedule Performance Index (SPI) | A measure of how efficiently a construction project is progressing compared to the planned schedule. It helps assess schedule performance. Example: An SPI value above 1 indicates that the construction project is ahead of schedule, while a value below 1 indicates a delay. Cross-reference: Refer to Earned Value for a measure of the value of work completed in a construction project.

Scope | The detailed description of what needs to be done in a construction project. It outlines the project's boundaries and objectives. Example: The project scope defines the specific tasks, deliverables, and objectives of the construction project. Cross-reference: See also Stakeholder for anyone affected by or affecting the construction project.

Stakeholder | Anyone who is affected by or can affect a construction project. It includes clients, workers, local communities, and others. Example: Stakeholders in a construction project may include the project owner, local residents, contractors, and regulatory authorities. Cross-reference: Refer to Communication Plan for strategies outlining how information will be shared among everyone involved in the construction project.

Subcontractor | A specialized company or individual hired by the main contractor to perform specific tasks in a construction project. Subcontractors are experts in particular trades. Example: A subcontractor may be hired for electrical work, providing specialized expertise within the broader construction project. Cross-reference: See also Bid for an official offer from a construction company to do a specific project.

Team Building | Fostering cooperation and collaboration among the individuals working on a construction project. It ensures a positive working environment. Example: Team building activities enhance communication and cooperation among construction project team members, fostering a collaborative work environment. Cross-reference: Refer to Human Resources for the people involved in a construction project, including workers, managers, and support staff.

Time Management | Efficiently allocating and using time in a construction project to meet deadlines. It involves planning and prioritizing tasks. Example: Time management ensures that tasks are completed on schedule, preventing delays in the overall construction project. Cross-reference: See also Project Management Plan for a detailed document outlining how a construction project will be executed, monitored, and controlled.

Total Float | The amount of time that a task or project can be delayed without affecting the overall project timeline. It provides flexibility in scheduling. Example: Total float allows for some delay in non-critical tasks without impacting the construction project's completion date. Cross-reference: Refer to Float for the amount of time that a task or project can be delayed without affecting the overall project timeline.

Value Engineering | A systematic approach to optimize the value of a construction project by balancing cost, performance, and quality. It aims to achieve the best value for the money spent. Example: Value engineering may involve exploring alternative materials or construction methods to achieve cost savings without compromising quality. Cross-reference: See also Lean Construction for an approach that focuses on minimizing waste and maximizing efficiency.

Work Breakdown Structure (WBS) | A hierarchical breakdown of all tasks in a construction project. It helps in organizing and managing project activities. Example: The WBS breaks down the construction project into manageable components, providing a structured view of tasks and subtasks. Cross-reference: Refer to Network Diagram for a visual representation of tasks and their relationships in a construction project.

Work Package | A small, manageable part of a construction project that can be assigned to a specific team or individual. It breaks down the project into more manageable components. Example: A work package may involve a specific phase of construction, such as foundation pouring, assigned to a specialized team. Cross-reference: See also Activity for a specific task or job in a construction project.

CHAPTER 3: QUALITY MANAGEMENT IN CONSTRUCTION

Acceptance Criteria | Criteria or standards set to determine whether a product or service meets required specifications and is accepted by the client or end-user. In construction, these criteria ensure that completed work aligns with specified quality and performance standards. (See also Quality Control)

Accreditation | Official recognition that an organization or individual has met specific standards and qualifications. In construction, accreditation may be related to certifications for quality management systems, demonstrating a commitment to industry-recognized standards. (Refer to ISO 9001 for international standards)

Audit | Systematic examination and evaluation of processes, procedures, or systems to ensure compliance with established standards and identify areas for improvement. Audits in construction assess whether project processes align with quality standards and provide insights for continuous improvement. (See also Continuous Improvement)

Benchmarking | Comparing performance metrics and practices against industry standards or best-in-class practices to identify areas for improvement. Construction companies use benchmarking to evaluate their project efficiency and identify opportunities for enhancement. (See also KPIs (Key Performance Indicators))

Change Control | A systematic process for managing changes to a project, ensuring that changes are documented, reviewed, and approved to prevent negative impacts on quality. In construction, change control addresses modifications to project plans to maintain quality standards. (Refer to Quality Assurance)

Compliance | Adherence to laws, regulations, and industry standards. In construction, compliance ensures that projects meet legal and regulatory requirements, covering aspects such as building codes and safety regulations. (See Quality Standards for criteria used to determine compliance)

Conformance | The degree to which a product, service, or process meets specified requirements or standards. In construction, conformance ensures that materials, workmanship, and processes align with the project's quality standards. (Refer to Quality Control)

Continuous Improvement | Ongoing efforts to enhance processes, products, or services incrementally. In construction, continuous improvement involves identifying and implementing improvements to achieve better quality outcomes in projects. (See also Six Sigma)

Defect | A flaw or imperfection in a product or service that deviates from the specified requirements. In construction, defects may include structural flaws, finishing errors, or deviations from architectural plans that require corrective action. (Refer to Root Cause Analysis)

Design Review | Evaluation of the design documentation to ensure it meets quality and performance requirements before construction begins. Design reviews in construction identify potential flaws in architectural plans, preventing issues during the construction phase. (See also Quality Assurance)

Inspection | A systematic examination of materials, components, or construction work to ensure they meet specified requirements. Inspections in construction involve checking materials, workmanship, and adherence to design specifications. (Refer to Quality Control)

ISO 9001 | An international standard for quality management systems that provides a framework for organizations to establish and maintain effective quality processes. Implementation of ISO 9001 in construction demonstrates a commitment to maintaining high-quality construction practices. (Refer to Quality Management System (QMS))

Non-Conformance | Failure of a product, service, or process to meet specified requirements or standards. In construction, non-conformance may involve a failure to meet building code requirements, necessitating corrective action. (See Quality Control for activities ensuring that construction processes meet specified requirements)

Performance Standards | Established criteria for measuring the effectiveness and efficiency of processes, products, or services. In construction, performance standards may include acceptable levels of material strength, project completion times, and budget adherence. (Refer to KPIs (Key Performance Indicators))

Quality Assurance | A set of activities and processes designed to ensure that project processes are implemented correctly to meet quality standards. Quality assurance in construction involves regular audits, reviews, and process checks to ensure compliance with quality requirements. (See also Design Review)

Quality Control | Activities and processes used to monitor and verify that construction processes and deliverables meet specified requirements. Quality control in construction involves inspections, testing, and verification to identify and correct defects. (Refer to Acceptance Criteria)

Quality Management System (QMS) | A comprehensive framework that provides a structured approach to managing an organization's processes to meet quality objectives. Implementation of a QMS in construction ensures consistency and adherence to quality standards across projects. (Refer to ISO 9001)

Quality Manual | A document that outlines the quality policies, procedures, and processes of an organization. A construction company's quality manual may specify procedures for materials testing, construction methods, and project documentation. (Refer to Quality Management System (QMS))

Quality Plan | A document that describes the quality standards, processes, and activities to be implemented in a construction project. A quality plan for a construction project may outline inspection schedules, testing protocols, and project milestones. (See Quality Standards)

Quality Policy | A statement that articulates an organization's commitment to quality and customer satisfaction. A construction company's quality policy may emphasize a commitment to delivering projects on time, within budget, and to specified quality standards. (See Quality Standards)

Quality Standards | Criteria used to determine whether a product, service, or process meets specified requirements. Quality standards in construction may include material strength, building code compliance, and adherence to project timelines. (Refer to Compliance)

Quality Surveillance | Ongoing monitoring and oversight activities to ensure that construction processes and deliverables meet specified quality requirements. Quality surveillance in construction may involve regular site visits, inspections, and reviews to monitor construction activities. (Refer to Quality Control)

Root Cause Analysis | A systematic process for identifying the underlying causes of problems or defects to implement effective corrective actions. Root cause analysis in construction may be used to identify the reasons behind schedule delays or material failures. (Refer to Continuous Improvement)

Six Sigma | A set of techniques and tools for process improvement, aiming to reduce defects and variability in processes. In construction, Six Sigma methodologies may be applied to improve efficiency, reduce errors, and enhance overall project quality. (See also Continuous Improvement)

Standard Operating Procedures (SOP) | Established procedures for performing repetitive tasks or activities to ensure consistency and quality. Standard operating procedures in construction may outline step-by-step processes for safety checks, material handling, or equipment operation. (Refer to Quality Assurance)

Statistical Process Control (SPC) | A statistical method for monitoring and controlling processes to ensure they operate within specified quality limits. Statistical process control in construction may involve monitoring variations in project timelines, costs, and material. (See Continuous Improvement)

CHAPTER 4: CONSTRUCTION SAFETY AND HEALTH

Accident Prevention Plan (APP) | An Accident Prevention Plan (APP) is a crucial document in construction safety, outlining strategies and protocols to mitigate the risk of accidents and enhance worker safety on construction sites. It provides a systematic approach to identify potential hazards, specifies preventive measures, and establishes emergency response procedures. The APP ensures that all stakeholders, including workers, supervisors, and management, are aware of safety protocols and contribute to maintaining a secure work environment. (Refer to Safety Audit for regular assessments of APP effectiveness.)

Asbestos | Asbestos is a naturally occurring fibrous mineral that was extensively used in construction materials for its insulation and fire-resistant properties. However, prolonged exposure to asbestos fibers poses severe health risks, including lung diseases such as asbestosis and mesothelioma. Proper handling, removal, and disposal of asbestos-containing materials are essential to prevent health hazards on construction sites. (See also Hazard Communication (HazCom) for guidelines on communicating asbestos-related risks.)

Behavior-Based Safety (BBS) | Behavior-Based Safety (BBS) is an approach that focuses on observing and modifying worker behavior to improve safety outcomes on construction sites. It recognizes that individual behaviors influence safety performance and emphasizes positive reinforcement, education, and communication to create a safety-conscious culture. BBS programs involve regular observations, feedback sessions, and continuous improvement initiatives. (Refer to Continuous Improvement for the ongoing enhancement of BBS programs.)

Confined Space | Confined spaces in construction pose unique safety challenges due to limited access and egress, as well as potential hazards like poor ventilation or toxic atmospheres. A confined space may include tanks, tunnels, or manholes. Implementing proper safety measures, such as confined space permits, atmospheric monitoring, and specialized training, is crucial to ensure the safety of workers entering and working in confined spaces. (See Emergency Evacuation Plan for preparedness in confined space emergencies.)

Emergency Evacuation Plan | An Emergency Evacuation Plan is a detailed strategy outlining procedures to safely evacuate a construction site in case of emergencies such as fires, natural disasters, or hazardous material spills. The plan includes evacuation routes, assembly points, communication protocols, and roles and responsibilities for site personnel. Regular drills and reviews are essential to ensure the effectiveness of the plan and the swift and safe evacuation of all workers. (See also Safety Audit for assessing and refining emergency preparedness.)

Fall Protection | Fall Protection encompasses a range of systems and measures designed to prevent falls from elevated surfaces, a significant hazard in construction. This includes the use of guardrails, safety nets, personal fall arrest systems, and other engineered solutions. Fall protection planning, proper equipment use, and worker training are essential components to mitigate the risk of falls and ensure worker safety. (Refer to

Job Hazard Analysis (JHA) for identifying fall hazards associated with specific tasks.)

Hazard Communication (HazCom) | Hazard Communication, often referred to as HazCom, is a standard ensuring that information about chemical hazards in the workplace is communicated to workers. This includes labeling of chemical containers, safety data sheets (SDS), and employee training. In construction, HazCom is critical for safeguarding workers against potential health hazards associated with exposure to chemicals commonly used in construction materials and processes. (See Material Safety Data Sheet (MSDS) for detailed information on chemical hazards.)

Incident Investigation | Incident Investigation is a systematic process conducted after a workplace incident to identify its causes and implement corrective actions to prevent recurrence. In construction, incidents may include accidents, injuries, or near misses. Conducting thorough investigations helps uncover root causes, contributing factors, and areas for improvement in safety protocols. (Refer to Root Cause Analysis for a detailed examination of underlying causes.)

Job Hazard Analysis (JHA) | Job Hazard Analysis (JHA) is a systematic process for identifying and evaluating potential hazards associated with specific job tasks in construction. It involves breaking down a job into individual steps, identifying potential hazards, and determining preventive measures to ensure worker safety. JHA is a proactive approach to prevent accidents and injuries by addressing specific job-related risks. (See Behavior-Based Safety (BBS) for reinforcing safe behaviors identified in JHA.)

Ladder Safety | Ladder Safety in construction involves guidelines and practices to ensure the safe use of ladders, a common tool for accessing elevated work areas. Proper ladder selection, inspection, placement, and use are essential to prevent falls and injuries. Worker training on ladder safety protocols and the use of additional fall protection measures when necessary contribute to maintaining a safe construction environment. (See Fall Protection for a broader approach to preventing falls from elevated surfaces.)

Material Safety Data Sheet (MSDS) | A Material Safety Data Sheet (MSDS) is a comprehensive document that provides information on the properties and hazards of chemicals used in the workplace. In construction, MSDS is essential for ensuring that workers are informed about the potential risks associated with the use, storage, and handling of various construction materials. MSDS includes details such as chemical composition, physical properties, health hazards, and emergency response procedures. (Refer to Hazard Communication (HazCom) for guidelines on communicating chemical hazards.)

Noise Control | Noise Control in construction involves measures and strategies to minimize excessive noise levels, protecting workers from potential hearing damage. This includes the use of noise barriers, hearing protection, equipment maintenance, and work scheduling to reduce noise exposure. Adhering to noise control practices ensures a safer and healthier work environment. (See Ventilation for a broader approach to environmental control on construction sites.)

Occupational Safety and Health Administration (OSHA) | The Occupational Safety and Health Administration (OSHA) is a federal agency in the United States responsible for setting and enforcing safety and health regulations in the workplace. OSHA regulations are instrumental in promoting and maintaining safe working conditions in the construction industry, covering areas such as fall protection, hazard communication, and personal protective equipment. (Refer to Safety Audit for regular assessments of OSHA compliance.)

Personal Protective Equipment (PPE) | Personal Protective

Equipment (PPE) comprises equipment worn by workers to minimize exposure to workplace hazards. In construction, PPE includes safety helmets, gloves, goggles, respiratory protection, and other specialized gear. Proper selection, use, and maintenance of PPE are critical to ensuring worker safety on construction sites. (See Fall Protection for a specific type of PPE related to preventing falls.)

Quick-Acting Clamps | Quick-Acting Clamps are devices used in construction for rapidly securing materials or components in place. They play a vital role in ensuring efficient and safe construction practices by providing quick and reliable methods for holding materials during various tasks. Proper training and usage guidelines contribute to the effective and safe use of quick-acting clamps. (See Toolbox Talks for informal safety discussions addressing the proper use of construction tools.)

Respiratory Protection | Respiratory Protection involves measures and equipment to protect workers from inhaling hazardous substances such as dust, fumes, and gases. In construction, where exposure to airborne contaminants is common, respiratory protection includes the use of respirators and proper training on their selection, use, and maintenance. Adhering to respiratory protection protocols is essential for maintaining a safe work environment. (See Material Safety Data Sheet (MSDS) for information on respiratory hazards associated with specific construction materials.)

Safety Audit | A Safety Audit is a systematic examination of a construction site's safety practices and conditions to ensure compliance with safety regulations and standards. Conducted regularly, safety audits help identify areas for improvement, assess the effectiveness of safety programs, and enhance overall safety performance. Safety audits may cover various aspects, including emergency preparedness, hazard communication, and the use of personal protective equipment. (Refer to Emergency Evacuation Plan for specific safety audit considerations related to emergency procedures.)

Toolbox Talks | Toolbox Talks are informal safety meetings held on construction sites to discuss specific safety issues, share information, and reinforce safe work practices. These talks typically involve workers and supervisors and provide a platform for addressing site-specific hazards, sharing lessons learned, and promoting a safety-conscious culture. Consistent and engaging toolbox talks contribute to ongoing safety awareness and education. (See Quick-Acting Clamps for a specific tool-related safety discussion.)

Unsafe Act | An Unsafe Act refers to any action that violates established safety procedures and increases the risk of accidents and injuries on a construction site. Identifying and addressing unsafe acts are crucial components of construction safety programs. This includes promoting a culture of accountability, providing adequate training, and encouraging open communication to prevent unsafe behaviors. (See Behavior-Based Safety (BBS) for a broader approach to reinforcing safe behaviors.)

Ventilation | Ventilation in construction involves the provision of fresh air to confined spaces or areas with poor air quality, ensuring a safe and breathable work environment. Proper ventilation systems, including natural and mechanical ventilation, help control exposure to airborne contaminants and contribute to overall worker health and safety. (Refer to Noise Control for a specific environmental control measure related to noise.)

Welding Safety | Welding Safety encompasses guidelines and practices to ensure the safe operation of welding equipment and protect workers from hazards associated with welding processes. This includes proper training, the use of personal protective equipment, and adherence to safety protocols such as ventilation requirements. Welding safety is crucial in preventing injuries and health risks associated with welding activities. (See Respiratory Protection for a specific focus on protecting against airborne contaminants in welding.)

X-Ray Inspection | X-Ray Inspection involves the use of X-ray technology to inspect structural components in construction projects. This non-destructive testing method allows for the detection of flaws, defects, or inconsistencies in materials without compromising their integrity. Proper safety measures, including radiation protection, are essential during X-ray inspections to safeguard the health of workers and the public. (See Emergency Evacuation Plan for considerations related to emergencies during X-ray inspections.)

Young Worker Safety | Young Worker Safety initiatives and practices are aimed at protecting the safety and health of individuals entering the construction industry at a young age. These efforts involve specialized training, mentorship programs, and age-appropriate safety protocols to address the unique challenges and risks faced by young workers in construction. (See Behavior-Based Safety (BBS) for reinforcing safe behaviors among young workers.)

Zero Accident Policy | A Zero Accident Policy reflects a commitment to maintaining a workplace environment where accidents and injuries are reduced to zero through comprehensive safety measures. This policy emphasizes a proactive approach to hazard identification, training, and continuous improvement to create a culture of safety on construction sites. (Refer to Safety Audit for ongoing assessments of a zero accident policy's effectiveness.)

CHAPTER 5:
CONSTRUCTION LAW
AND CONTRACTS

Adjudication | Adjudication in the context of construction law refers to a formal dispute resolution process, often used in the UK, where an independent adjudicator makes a swift decision on a construction dispute. Adjudication is designed to provide a quick resolution to issues such as payment disputes or disagreements over variations in construction contracts. The adjudicator's decision is typically binding temporarily and can be challenged through other dispute resolution mechanisms.

Alternative Dispute Resolution (ADR) | Alternative Dispute Resolution (ADR) encompasses various methods, including mediation and arbitration, used to resolve construction disputes outside of traditional court litigation. ADR provides parties with a more flexible and often faster means of resolving conflicts. It is commonly included in construction contracts as a mandatory step before pursuing formal legal action.

Arbitration | Arbitration is a form of dispute resolution in construction contracts where parties present their case to an impartial arbitrator or panel. The arbitrator's decision is usually binding and enforceable. Arbitration offers a more private and streamlined alternative to traditional litigation, and the process can be tailored to the specific needs of the parties involved.

Breach of Contract | Breach of Contract occurs when one party fails to fulfill its contractual obligations without a legal excuse. In construction contracts, this might involve failure to meet project deadlines, inadequate workmanship, or non-compliance with specifications. Remedies for a breach of contract may include monetary damages or specific performance, depending on the terms of the contract.

Change Order | A Change Order is a written agreement between parties in a construction contract that modifies the original terms and conditions. It may involve changes in project scope, specifications, or timelines. Change orders are crucial for documenting alterations to the contract and ensuring that both parties agree to the modifications.

Claim | In construction, a Claim is a demand by one party seeking compensation or relief for issues such as delays, disruptions, or additional costs encountered during the project. Claims are often formalized in accordance with the dispute resolution process outlined in the construction contract.

Collateral Warranty | A Collateral Warranty is a legal document in construction contracts that creates a direct contractual relationship between a third party (such as a design professional or subcontractor) and the client or main contractor. It provides the third party with contractual obligations and liabilities, independent of the main contract.

Common Law | Common Law refers to legal principles established through court decisions rather than statutes. In construction law, common law plays a significant role in shaping legal standards and precedents related to contracts, negligence, and other legal issues.

Conditions of Contract | Conditions of Contract are the specific terms and provisions governing the rights and responsibilities of the parties involved in a construction contract. These conditions outline the contractual obligations, dispute resolution mechanisms, payment terms, and other critical aspects.

Contract Documents | Contract Documents encompass all the written, graphic, and pictorial documents that form a construction contract. This includes the contract itself, drawings, specifications, change orders, and other relevant materials. Clear and comprehensive contract documents are essential for avoiding misunderstandings and disputes.

Contractor | A Contractor is an individual or entity engaged to perform construction work under a contract. Contractors can be general contractors overseeing the entire project or specialty contractors responsible for specific aspects such as plumbing or electrical work.

Defect | In construction, a Defect refers to any imperfection or deficiency in the work that deviates from the agreed-upon specifications or industry standards. Defects can manifest in materials, workmanship, or design and may result in the need for corrective measures.

Design-Build | Design-Build is a project delivery method in which a single entity, often a design-build contractor, is responsible for both the design and construction phases. This integrated approach is intended to streamline communication, reduce risks, and enhance project efficiency.

Due Diligence | Due Diligence involves the careful investigation and evaluation of relevant factors before entering into a construction contract or transaction. It includes assessing legal, financial, and technical aspects to make informed decisions and mitigate potential risks.

Earnest Money | Earnest Money, also known as a bid bond, is a financial commitment provided by a bidder during the bidding process to demonstrate serious intent to enter into a contract if awarded. If the bidder fails to fulfill their obligations, the earnest money may be forfeited.

Estoppel | Estoppel is a legal principle preventing a party from asserting a claim or defense that is inconsistent with their previous conduct or statements. In construction contracts, estoppel may arise when one party relies on the representations or actions of another to their detriment.

Force Majeure | Force Majeure refers to unforeseen circumstances or events beyond the control of the parties that may excuse non-performance of contractual obligations. Common force majeure events in construction include natural disasters, wars, and government actions.

Indemnity | Indemnity is a contractual provision where one party agrees to compensate the other for losses, damages, or liabilities arising from specified events. Construction contracts often include indemnity clauses to allocate risks between parties.

Liquidated Damages | Liquidated Damages are predetermined and agreed-upon damages specified in a construction contract to compensate the injured party for specific breaches, such as delays in project completion. These damages are intended to reflect a reasonable estimate of the actual harm caused.

Mechanic's Lien | A Mechanic's Lien is a legal claim filed by contractors, subcontractors, or suppliers against a property's title to secure payment for labor or materials provided in a construction project. Mechanic's liens are tools to protect the rights of those contributing to the construction process.

Mediation | Mediation is a form of alternative dispute resolution (ADR) where an impartial third party, the mediator, facilitates negotiations between parties to help them reach a mutually acceptable resolution. Mediation is non-binding, and the parties retain control over the outcome.

Negligence | Negligence, in the context of construction law, involves the failure to exercise reasonable care, resulting in harm or damage to others. Claims of negligence may arise from design errors, construction defects, or inadequate supervision.

Novation | Novation is the substitution of one party in a contract with the consent of all involved parties. In construction, novation may occur when a new contractor or subcontractor assumes the obligations and rights of the original party.

Preliminary Notice | A Preliminary Notice, also known as a notice to owner or notice of furnishing, is a formal document sent by subcontractors or suppliers to notify property owners, contractors, and other relevant parties of their involvement in a construction project. Preliminary notices help protect the sender's lien rights.

Quantum Meruit | Quantum Meruit is a legal doctrine that allows a party to recover a reasonable value for goods or services provided, even in the absence of a formal contract. It applies when one party confers a benefit on another, and fairness dictates compensation.

Retainage | Retainage, also known as retention, is a portion of a contractor's payment withheld by the client until the completion of the construction project. Retainage serves as a form of security to ensure the contractor fulfills all contractual obligations.

Scope of Work | The Scope of Work defines the specific tasks, responsibilities, and deliverables associated with a construction project. It outlines the project's boundaries and clarifies what work is included and excluded from the contract.

Statute of Limitations | The Statute of Limitations sets the maximum time period within which legal proceedings can be initiated. In construction law, this refers to the timeframe during which parties can file claims or lawsuits related to construction defects, contract breaches, or other issues.

Subcontractor | A Subcontractor is an individual or entity hired by the main contractor to perform specific tasks or provide services as part of a construction project. Subcontractors are typically specialists in particular trades.

Surety Bond | A Surety Bond is a financial guarantee provided by a third party (the surety) to ensure that a contractor fulfills their contractual obligations. If the contractor fails to perform, the surety may be required to compensate the project owner.

Tort | Tort refers to a civil wrong or injury inflicted on one party by another, leading to legal liability. In construction law, tort claims may arise from negligence, personal injury, or property damage.

Waiver | A Waiver is the intentional relinquishment of a known right, often evidenced by a written or explicit statement. In construction contracts, waivers may be used to modify or release certain contractual rights.

CHAPTER 6: SUSTAINABLE CONSTRUCTION PRACTICES

Biodiversity | Biodiversity in sustainable construction refers to the variety of plant and animal life within a specific ecosystem. Sustainable practices aim to preserve and enhance biodiversity by incorporating green spaces, native plantings, and wildlife habitats into construction projects. For example, green roofs with native vegetation can promote biodiversity in urban environments. (See Green Building Certification for sustainability standards that may include biodiversity criteria.)

Building Information Modeling (BIM) | Building Information Modeling (BIM) is a digital representation of the physical and functional characteristics of a building. In sustainable construction, BIM facilitates the efficient design, construction, and operation of environmentally friendly buildings. BIM allows stakeholders to visualize and analyze various aspects, such as energy performance and material use, throughout the building's lifecycle. (Refer to Life Cycle Assessment (LCA) for a holistic approach to assessing environmental impacts.)

Carbon Footprint | The Carbon Footprint of a construction project represents the total amount of greenhouse gas emissions, typically measured in carbon dioxide equivalents, associated with its activities. Sustainable construction aims to minimize carbon footprints by using energy-efficient materials, reducing transportation emissions, and incorporating renewable energy sources. For instance, choosing locally sourced materials can lower transportation-related carbon emissions. (See Net Zero Energy Building for a goal of achieving zero net energy consumption.)

Circular Economy | The Circular Economy is an approach that emphasizes minimizing waste and maximizing the reuse, recycling, and repurposing of materials. In sustainable construction, adopting circular economy principles involves designing buildings for disassembly, recycling construction waste, and promoting the reuse of salvaged materials. For example, repurposing reclaimed wood for new construction projects aligns with circular economy principles. (Explore Recycling as a key component of circular economy practices.)

Daylighting | Daylighting is the intentional use of natural light to illuminate interior spaces, reducing the need for artificial lighting. Sustainable construction incorporates daylighting strategies through the design of windows, skylights, and reflective surfaces. For instance, strategically placing windows and using light-colored materials can optimize natural daylight and enhance energy efficiency. (Consider Indoor Air Quality as daylighting contributes to creating healthier indoor environments.)

Eco-Friendly Materials | Eco-Friendly Materials in sustainable construction are those that have a reduced impact on the environment throughout their lifecycle. Examples include recycled steel, bamboo, reclaimed wood, and low-VOC (volatile organic compound) paints. Choosing eco-friendly materials supports sustainable practices by reducing resource depletion and minimizing environmental harm. (See LEED (Leadership in Energy and Environmental Design) for green building certification criteria related to material selection.)

Energy Efficiency | Energy Efficiency in sustainable construction involves designing and constructing buildings that optimize energy use to minimize waste. Sustainable practices include incorporating energy-efficient appliances, lighting, insulation, and HVAC systems. For example, installing energy-efficient windows can reduce the need for artificial heating and cooling. (Explore Renewable Energy as part of an integrated approach to achieving energy efficiency.)

Geothermal Heating and Cooling | Geothermal Heating and Cooling systems utilize the Earth's stable temperature to regulate indoor climate conditions. In sustainable construction, geothermal systems reduce reliance on traditional heating and cooling methods, contributing to energy efficiency. For instance, geothermal heat pumps can efficiently transfer heat between the ground and a building for heating or cooling purposes. (Refer to Renewable Energy for an overview of sustainable energy sources.)

Green Building Certification | Green Building Certification, such as LEED (Leadership in Energy and Environmental Design), is a third-party verification process that assesses a building's sustainability performance. Sustainable construction projects may pursue certification to demonstrate adherence to environmental standards. For example, achieving LEED certification requires meeting criteria related to energy efficiency, water conservation, and indoor environmental quality. (Cross-reference with Biodiversity to emphasize the multifaceted nature of sustainability certification.)

Indoor Air Quality | Indoor Air Quality focuses on creating healthy and comfortable indoor environments by minimizing pollutants and maximizing ventilation. Sustainable construction practices include selecting low-VOC materials, providing adequate ventilation, and designing spaces that optimize air circulation. For instance, incorporating daylighting strategies can enhance indoor air quality by reducing the reliance on artificial lighting and associated pollutants. (Consider Daylighting as part of a holistic approach to sustainable design.)

LEED (Leadership in Energy and Environmental Design) | LEED is a widely recognized green building certification program that evaluates the environmental performance of buildings. Sustainable construction projects seeking LEED certification must meet specific criteria related to energy efficiency, water conservation, materials selection, and indoor environmental quality. For example, achieving LEED certification may involve incorporating renewable energy sources and implementing sustainable water management practices. (Explore Eco-Friendly Materials for a key aspect of LEED certification.)

Life Cycle Assessment (LCA) | Life Cycle Assessment is a comprehensive evaluation of the environmental impacts of a product or process throughout its entire lifecycle. In sustainable construction, LCA assesses factors such as resource use, energy consumption, and emissions. For instance, using LCA helps identify the most environmentally friendly materials and construction methods. (Cross-reference with Building Information Modeling (BIM) for a digital tool that aids in analyzing life cycle impacts.)

Net Zero Energy Building | A Net Zero Energy Building generates as much energy as it consumes over the course of a year. Sustainable construction practices to achieve net-zero energy include incorporating renewable energy sources, optimizing insulation, and using energy-efficient systems. For example, installing solar panels on a building's roof can contribute to achieving a net-zero energy balance. (See Carbon Footprint for an interconnected goal of minimizing overall environmental impact.)

Passive Design | Passive Design in sustainable construction involves designing buildings to naturally optimize heating, cooling, and lighting without relying on mechanical systems. Sustainable practices include strategic building orientation, proper insulation, and the use of thermal mass. For example, designing windows to maximize natural light and heat in winter while providing shade in summer is a key element of passive design. (Consider Daylighting as part of an integrated passive design strategy.)

Rainwater Harvesting | Rainwater Harvesting involves collecting and storing rainwater for later use, reducing the demand on traditional water sources. In sustainable construction, rainwater harvesting systems may include rooftop collection and storage tanks. For instance, harvested rainwater can be used for irrigation, flushing toilets, or other non-potable applications, contributing to water conservation efforts. (Cross-reference with Water Conservation for a broader perspective on sustainable water management.)

Recycling | Recycling in sustainable construction involves processing and reusing materials from demolished or deconstructed buildings. Sustainable practices include recycling concrete, steel, wood, and other materials to reduce waste and conserve resources. For example, crushed recycled concrete can be used as aggregate in new construction projects. (See Circular Economy for a broader concept that encompasses recycling within a holistic approach.)

Renewable Energy | Renewable Energy sources, such as solar, wind, and geothermal power, are integral to sustainable construction. Sustainable projects prioritize the use of renewable energy to reduce reliance on non-renewable sources and minimize environmental impact. For example, installing solar panels on a building's roof generates clean and sustainable electricity. (Explore Geothermal Heating and Cooling for a specific application of renewable energy in sustainable construction.)

Sustainable Architecture | Sustainable Architecture emphasizes environmentally conscious design principles that minimize the ecological impact of buildings. Sustainable architects consider factors such as energy efficiency, material selection, and site impact. For example, incorporating passive design strategies and eco-friendly materials are key elements of sustainable architectural practices. (Cross-reference with Building Information Modeling (BIM) for a digital tool that aids in visualizing and implementing sustainable architectural concepts.)

Sustainable Development | Sustainable Development in construction involves creating projects that meet present needs without compromising the ability of future generations to meet their own needs. Sustainable development considers economic, social, and environmental factors. For instance, a sustainable development project may integrate green spaces, promote energy efficiency, and prioritize community well-being. (See Triple Bottom Line for a broader concept that evaluates sustainability from economic, social, and environmental perspectives.)

Sustainable Urban Planning | Sustainable Urban Planning focuses on creating environmentally friendly and livable urban environments. In construction, sustainable urban planning integrates green spaces, public transportation, and energy-efficient infrastructure. For example, designing walkable neighborhoods with access to public transit promotes sustainability in urban areas. (Cross-reference with Urban Heat Island (UHI) Effect to address challenges related to urban heat.)

Triple Bottom Line | The Triple Bottom Line evaluates sustainability by considering economic, social, and environmental factors. In construction, the triple bottom line assesses projects based on their impact on profit, people, and the planet. For example, a sustainable construction project may enhance energy efficiency, support local communities, and reduce environmental degradation. (See Sustainable Development for an application of the triple bottom line in construction.)

Upcycling | Upcycling is the process of transforming waste materials or unwanted products into new materials or products of higher value. In sustainable construction, upcycling may involve repurposing salvaged materials for new building projects. For instance, using reclaimed wood from old structures for flooring or furniture represents an upcycling practice. (Cross-reference with Circular Economy for an overarching concept that includes upcycling within sustainable material practices.)

Urban Heat Island (UHI) Effect | The Urban Heat Island (UHI) Effect refers to the elevated temperatures in urban areas compared to their rural surroundings. In sustainable construction, addressing the UHI effect involves incorporating green spaces, cool roofs, and reflective surfaces. For example, planting trees and creating green roofs can mitigate the UHI effect by providing shade and reducing heat absorption. (Explore Sustainable Urban Planning for a broader approach to mitigating urban environmental challenges.)

Water Conservation | Water Conservation in sustainable construction involves implementing practices to reduce water usage and promote responsible water management. Sustainable projects may include efficient irrigation systems, low-flow fixtures, and rainwater harvesting. For example, designing landscapes with native and drought-tolerant plants minimizes the need for excessive water use. (See Rainwater Harvesting for a specific practice related to water conservation in construction.)

Wind Power | Wind Power is a renewable energy source

harnessed to generate electricity. In sustainable construction, incorporating wind power may involve installing wind turbines to supplement a building's energy needs. For instance, on-site wind turbines can contribute to a project's overall energy efficiency and sustainability. (Explore Renewable Energy for a broader perspective on incorporating various renewable sources in construction.)

PART II: CONSTRUCTION MATERIALS AND METHODS

CHAPTER 7: PROPERTIES AND APPLICATIONS OF CONSTRUCTION MATERIALS

Adhesion | Adhesion in construction materials refers to the bonding between different materials, ensuring their attachment. This property is crucial for the durability and stability of structures. For example, in the application of tiles, proper adhesion between the tile and the substrate is essential to prevent detachment over time. (Refer to Reinforced Concrete for instances where adhesion is critical between concrete and reinforcing materials.)

Aggregate | Aggregate in construction materials consists of granular particles used to form the skeleton of concrete or mortar. Examples include gravel, sand, or crushed stone. The selection of aggregates impacts the strength and durability of concrete. For instance, using well-graded aggregates enhances the workability and strength of concrete mixtures. (See Concrete for a comprehensive understanding of how aggregates contribute to concrete properties.)

Alkali-Silica Reaction (ASR) | Alkali-Silica Reaction is a chemical reaction in concrete between alkalis in the cement and certain types of silica minerals in aggregates, leading to expansion and cracking. Mitigating ASR is crucial for the durability of concrete structures. For example, using low-alkali cements or selecting ASR-resistant aggregates helps prevent deleterious effects. (Cross-reference with Concrete for the broader context of concrete-related phenomena.)

Brick | Brick is a traditional construction material made from clay or shale that undergoes a firing process. Bricks are versatile and used in various construction applications, such as walls, facades, and pavements. For instance, bricks are renowned for their durability and aesthetic appeal in building construction. (Explore Masonry for a broader understanding of brick and other masonry materials.)

Cement | Cement is a binding agent in concrete that provides cohesion and strength. Ordinary Portland Cement (OPC) is a common type used in construction. For example, in the production of concrete, cement binds aggregates and water to form a durable and versatile construction material. (Refer to Concrete for an in-depth analysis of the role of cement in concrete mixtures.)

Concrete | Concrete is a composite construction material consisting of cement, aggregates, water, and admixtures. It is widely used in various applications, including foundations, pavements, and structures. For instance, reinforced concrete combines concrete's compressive strength with the tensile strength of reinforcing materials for enhanced structural performance. (See Reinforced Concrete for a specific application of concrete in structural engineering.)

Corrosion | Corrosion in construction materials refers to the deterioration of metals due to chemical reactions with the environment. Preventing corrosion is vital for maintaining the integrity of structures. For example, applying corrosion-resistant coatings or using stainless steel reinforcement helps protect against corrosion in concrete structures. (Cross-reference with Steel for an in-depth look at corrosion in steel.)

Durability | Durability is the ability of construction materials to withstand environmental conditions and maintain their integrity over time. It is a crucial factor in material selection for long-lasting structures. For instance, in bridge construction, selecting durable materials ensures the longevity of the structure despite exposure to harsh weather conditions. (Explore Reinforced Concrete for a comprehensive understanding of how durability influences material selection.)

Engineered Wood | Engineered Wood comprises wood products manufactured by binding or fixing strands, particles, fibers, or veneers together. Examples include plywood and laminated veneer lumber. For example, engineered wood products offer enhanced structural properties and are used in various applications, such as flooring and framing. (See Wood Preservation for considerations related to preserving engineered wood.)

Fiber-Reinforced Concrete (FRC) | Fiber-Reinforced Concrete incorporates fibers, such as steel or synthetic materials, to enhance its structural performance. FRC improves tensile strength and crack resistance. For instance, in tunnel construction, FRC is utilized to enhance the durability and resilience of concrete linings. (Cross-reference with Concrete for a broader understanding of fiber-reinforced concrete applications.)

Fire Resistance | Fire Resistance in construction materials refers to their ability to withstand exposure to fire without compromising structural integrity. Materials like fire-resistant coatings and gypsum boards contribute to fire protection in buildings. For example, in skyscraper construction, fire-resistant materials are essential for ensuring the safety of occupants and preventing structural collapse during a fire. (See Steel for considerations related to fire resistance in structural steel.)

Glass | Glass is a versatile construction material used for windows, facades, and decorative elements. Its transparency allows natural light into buildings. For example, in modern architectural designs, glass curtain walls are employed to maximize daylighting and create visually appealing facades. (Explore Insulation for considerations related to thermal and acoustic properties of glass.)

Insulation | Insulation in construction materials is crucial for regulating temperature and sound within structures. Examples include fiberglass and foam board. For instance, in residential construction, proper insulation contributes to energy efficiency by minimizing heat transfer and reducing the need for heating and cooling. (See Glass for considerations related to glass as an insulating material.)

Lime Mortar | Lime Mortar is a traditional mortar composed of lime, sand, and water. It is used in masonry construction for its workability and compatibility with historic structures. For example, in the restoration of historical buildings, lime mortar is preferred to maintain the authenticity and breathability of the structure. (Cross-reference with Masonry for insights into the use of lime mortar in traditional masonry.)

Masonry | Masonry involves the use of building units, such as bricks or stones, and a binding material like mortar. It is a versatile construction method used in walls, facades, and historical structures. For instance, in the construction of load-bearing walls, masonry provides strength and durability. (Explore Brick and Lime Mortar for specific applications and considerations within masonry construction.)

Metal | Metal, such as steel and aluminum, is widely used in construction for its strength and versatility. Metals find applications in structural elements like beams and columns. For example, in high-rise building construction, steel framing provides the necessary strength to support the structure. (See Corrosion for considerations related to preventing corrosion in metal construction elements.)

Plywood | Plywood is an engineered wood product consisting of thin layers of wood veneer glued together. It is used for its strength and dimensional stability. For instance, in the construction of formwork for concrete, plywood provides a smooth and durable surface. (Cross-reference with Engineered Wood for insights into the use of plywood as an engineered wood product.)

Polymers | Polymers in construction materials refer to synthetic substances with high molecular weights. They are used in various applications, including adhesives, sealants, and coatings. For example, in waterproofing applications, polymer-modified materials enhance the durability and resistance of construction elements to water infiltration. (See Adhesion for considerations related to the adhesion of polymers in construction.)

Rebar | Rebar, or reinforcing bar, is a steel bar used in reinforced concrete to improve its tensile strength. For example, in the construction of a reinforced concrete slab, rebar provides additional strength to resist bending and cracking. (Explore Concrete for insights into the role of rebar in enhancing the structural performance of concrete.)

Reinforced Concrete | Reinforced Concrete is a composite material consisting of concrete and embedded reinforcement, typically steel. It combines the compressive strength of concrete with the tensile strength of steel. For instance, in the construction of bridges, reinforced concrete structures offer a balance of strength and durability. (Cross-reference with Rebar for specific considerations related to reinforcing concrete.)

Steel | Steel is a versatile construction material known for its strength and ductility. It finds applications in structural elements, such as beams and columns. For example, in the construction of large-span structures like stadiums, steel framing provides the necessary support and allows for open and flexible designs. (See Corrosion for insights into preventing corrosion in structural steel.)

Stone | Stone is a natural construction material used for various applications, including cladding, flooring, and monuments. Its durability and aesthetic appeal make it a timeless choice. For instance, in the construction of historical landmarks, the use of natural stone contributes to the cultural and architectural significance of the structure. (Explore Masonry for considerations related to the use of stone in masonry construction.)

Stucco | Stucco is a traditional exterior plaster finish made from cement, sand, and lime. It is applied over a lath for decorative and protective purposes. For example, in residential construction, stucco provides a durable and visually appealing finish for exterior walls. (See Lime Mortar for insights into traditional mortar finishes in construction.)

Thermal Conductivity | Thermal Conductivity is the property of construction materials that determines their ability to conduct heat. Materials with low thermal conductivity are often used for insulation. For instance, in energy-efficient building construction, materials with low thermal conductivity help in maintaining comfortable indoor temperatures. (Explore Insulation for considerations related to thermal properties of construction materials.)

Timber | Timber, or wood, is a renewable construction material used for structural elements, flooring, and finishes. For example, in residential construction, timber framing provides a sustainable and aesthetically pleasing option for building structures. (See Wood Preservation for considerations related to preserving timber in construction.)

Water Absorption | Water Absorption is the capacity of construction materials to absorb and retain water. For example, in the selection of exterior cladding, materials with low water absorption are preferred to prevent water damage and ensure longevity. (See Durability for insights into how water absorption affects the durability of construction materials.)

Wood Preservation | Wood Preservation involves treating timber to enhance its resistance to decay, insects, and environmental conditions. Preserved wood is used in various construction applications. For instance, in outdoor structures like decks, using preserved wood ensures longevity and resistance to decay. (Cross-reference with Timber for considerations related to preserving wood as a construction material.)

CHAPTER 8: CONCRETE TECHNOLOGY AND MIX DESIGN

Admixture | Admixtures in concrete technology are additional ingredients added to the concrete mix to modify its properties. They can enhance workability, durability, or set time. For example, the addition of a water-reducing admixture can improve the workability of a concrete mix, allowing for easier placement and finishing. (Refer to Mix Design for insights into how admixtures are considered in designing concrete mixes.)

Air Entrainment | Air entrainment is the intentional incorporation of air bubbles into concrete to improve its resistance to freeze-thaw cycles. This is crucial in regions with cold climates. For instance, in the construction of outdoor structures in cold climates, air-entrained concrete helps prevent damage caused by the expansion of freezing water within the concrete. (Cross-reference with Cold Joint for considerations related to freezing and thawing in concrete.)

Alkali-Aggregate Reaction (AAR) | Alkali-Aggregate Reaction is a chemical reaction between alkalies in cement and certain minerals in aggregates, leading to expansion and cracking.

Mitigating AAR is essential for the durability of concrete structures. For example, using low-alkali cements or selecting AAR-resistant aggregates helps prevent deleterious effects. (See Concrete for a broader context of concrete-related phenomena.)

Bleeding | Bleeding in concrete occurs when water from the mix rises to the surface before the concrete sets. It can affect the surface finish and durability. For instance, in the construction of high-rise buildings, controlling bleeding is crucial to achieve a smooth and durable concrete finish. (Explore High-Performance Concrete (HPC) for considerations related to bleeding in specialized concrete mixes.)

Curing | Curing is the process of maintaining adequate moisture, temperature, and time to allow concrete to achieve its desired strength and durability. Proper curing is essential for preventing cracks and ensuring long-term performance. For example, in the construction of bridge decks, meticulous curing practices contribute to the durability of the concrete in aggressive environments. (Cross-reference with Shrinkage for insights into how curing influences shrinkage in concrete.)

Fly Ash | Fly Ash is a pozzolanic material obtained from the combustion of pulverized coal in power plants. It is used as a supplementary cementitious material in concrete. For instance, in sustainable construction, incorporating fly ash into concrete mixes enhances durability and reduces the environmental impact of cement production. (See Pozzolan for a broader understanding of pozzolanic materials in concrete.)

High-Performance Concrete (HPC) | High-Performance Concrete is a specialized type of concrete designed to exhibit enhanced strength, durability, and workability. It often includes additives like silica fume or superplasticizers. For example, in the construction of infrastructure projects, such as high-rise buildings or bridges, HPC is employed to meet specific performance requirements. (Refer to Bleeding for considerations related to bleeding in high-performance concrete mixes.)

Hot Weather Concreting | Hot Weather Concreting involves special considerations and precautions when placing and curing concrete in high-temperature conditions. For instance, in desert regions, where ambient temperatures can be high, adjusting mix designs and implementing rapid curing methods are essential to prevent issues like premature setting and reduced workability. (Explore Cold Joint for considerations related to temperature differentials in concrete placement.)

Low-Slump Concrete | Low-Slump Concrete refers to concrete mixes with minimal workability or flowability. It is often used in applications where high strength and stiffness are required. For example, in the construction of structural elements like columns, low-slump concrete allows for precise placement and consolidation without excessive deformation. (Cross-reference with Slump Test for insights into measuring the consistency of low-slump concrete.)

Mix Design | Mix Design is the process of proportioning concrete ingredients to achieve the desired properties. It involves considerations such as strength, workability, and durability. For instance, in the construction of a high-rise building, a mix design tailored to meet the specific requirements of the structure ensures optimal performance. (See Admixture for insights into how admixtures influence the mix design process.)

Pozzolan | Pozzolans are materials, such as fly ash or silica fume, that react with lime in the presence of water to form cementitious compounds. They are used as supplementary cementitious materials in concrete. For example, in sustainable construction practices, incorporating pozzolans into concrete mixes reduces the environmental impact of cement production and enhances long-term performance. (Refer to Fly Ash for considerations related to using fly ash as a pozzolan in concrete.)

Ready-Mix Concrete | Ready-Mix Concrete is pre-mixed concrete delivered to construction sites in a ready-to-use form. It offers convenience and consistent quality. For instance, in large-scale construction projects like highway infrastructure, the use of ready-mix concrete streamlines the construction process and ensures uniformity in concrete quality. (Explore Retarder for insights into additives used in ready-mix concrete to control setting time.)

Retarder | A Retarder is an admixture used in concrete to slow down the setting time. It is beneficial in scenarios where extended workability is required. For example, in the construction of complex concrete structures, the use of retarders allows for adequate placement and finishing before the concrete begins to set. (Cross-reference with Ready-Mix Concrete for considerations related to the use of retarders in ready-mix formulations.)

Segregation | Segregation in concrete occurs when the components of the mix separate, leading to uneven distribution of aggregates and water. Controlling segregation is crucial for achieving uniform concrete properties. For instance, in the construction of high-strength concrete elements, preventing segregation ensures consistent strength throughout the structure. (See Air Entrainment for considerations related to air-entrained concrete and resistance to segregation.)

Self-Compacting Concrete (SCC) | Self-Compacting Concrete is a highly flowable concrete that can spread and fill formwork under its own weight. It is especially useful in complex structures with congested reinforcement. For example, in the construction of intricate architectural elements, SCC facilitates efficient placement without the need for excessive vibration. (Refer to Vibration for insights into the role of vibration in traditional concrete placement.)

Shrinkage | Shrinkage in concrete refers to the reduction in volume as the concrete cures and dries. It can lead to cracking if not adequately controlled. For instance, in the construction of large concrete slabs, minimizing shrinkage is essential to prevent surface cracking and ensure long-term durability. (Explore Curing for considerations related to curing practices and their impact on shrinkage in concrete.)

Silica Fume | Silica Fume is a pozzolanic material that improves the properties of concrete, including strength and durability. It is a byproduct of silicon metal production. For example, in the construction of high-strength concrete elements like precast components, the addition of silica fume enhances the material's performance. (See High-Performance Concrete (HPC) for considerations related to using silica fume in high-performance concrete mixes.)

Slump Test | The Slump Test is a standard test to measure the consistency of fresh concrete. It involves assessing the slump or sag of a cone-shaped concrete sample. For instance, in quality control during concrete placement, performing slump tests ensures that the concrete meets the desired workability requirements. (Cross-reference with Low-Slump Concrete for considerations related to measuring consistency in low-slump concrete.)

Superplasticizer | A Superplasticizer is an admixture that increases the workability of concrete without significantly affecting water content. It is often used in high-performance and self-compacting concrete. For example, in the construction of complex architectural shapes, superplasticizers enable the placement of highly workable concrete without sacrificing strength. (Refer to High-Performance Concrete (HPC) for insights into using superplasticizers in high-performance concrete.)

Vibration | Vibration is a process used in concrete placement to eliminate air voids and ensure proper compaction. It is typically applied through vibrating tools. For instance, in the construction of reinforced concrete structures, vibration is crucial to achieve uniform distribution of concrete around reinforcement and eliminate potential voids. (See Self-Compacting Concrete (SCC) for considerations related to the self-leveling properties that reduce the need for external vibration.)

CHAPTER 9: STEEL CONSTRUCTION SYSTEMS AND DESIGN

Bearing Plate | A Bearing Plate is a steel plate used to distribute the load from a structural member, such as a column or beam, to the supporting foundation. For example, in the construction of a steel frame building, bearing plates are strategically placed at the base of columns to ensure uniform load transfer to the foundation. (See Column for considerations related to the load-carrying vertical members in steel construction.)

Bolted Connection | A Bolted Connection involves joining steel members using bolts. It provides flexibility during construction and facilitates disassembly if needed. For instance, in the assembly of steel trusses for a roof structure, bolted connections allow for efficient on-site construction and adjustments. (Cross-reference with Welded Connection for comparisons between bolted and welded joining methods.)

Built-Up Section | A Built-Up Section is a steel member created by combining individual steel shapes, such as plates and angles, to achieve a specific structural profile. For example, in the design of a steel beam with increased load-carrying capacity, a built-up section may be created by welding or bolting together multiple steel components. (Refer to Steel Frame for insights into the use of built-up sections in overall steel frame design.)

Cold-Formed Steel | Cold-Formed Steel refers to steel sections that are shaped at room temperature using processes such as rolling or pressing. It is commonly used in light-gauge structural elements. For instance, in residential construction, cold-formed steel studs are often employed in framing interior walls due to their lightweight and cost-effective nature. (Cross-reference with Shear Connection for considerations related to connecting cold-formed steel members.)

Column | A Column is a vertical load-bearing member in a steel frame structure that supports vertical loads from beams and other structural elements. For example, in the construction of a multi-story building, steel columns provide essential support to the floors above. (See Bearing Plate for insights into the role of bearing plates in distributing column loads to the foundation.)

Composite Construction | Composite Construction involves combining steel and concrete to create a structural system with enhanced strength and efficiency. For instance, in the construction of a composite floor system, steel beams are connected to a concrete slab to optimize load-carrying capabilities. (Cross-reference with Moment Connection for considerations related to creating strong connections in composite construction.)

Connection Design | Connection Design is the process of determining the type and specifications of connections between steel members to ensure structural integrity. For example, in the design of a steel truss, connection design considers factors like load transfer and stability to ensure the overall structural performance. (See Bolted Connection for insights into the design considerations of bolted connections in steel structures.)

Erection | Erection in steel construction refers to the process of assembling and installing structural steel members on-site. For instance, in the construction of a steel-framed industrial facility, the erection process involves cranes and specialized equipment to position and connect steel members according to the engineered design. (Cross-reference with Flange for insights into the role of flanges in the stability of erected steel structures.)

Flange | In steel construction, a Flange is the projecting edge of a structural shape, such as an I-beam, that provides resistance to bending. For example, in the design of a steel beam, the flange contributes significantly to the beam's load-carrying capacity. (Refer to Moment Frame for considerations related to using moment frames where flanges play a crucial role.)

Moment Connection | A Moment Connection is a type of steel connection that allows the transfer of significant moments or rotational forces between connected members. For instance, in the design of a steel moment frame for seismic resistance, moment connections ensure effective load transfer and stability. (See Composite Construction for insights into creating moment connections in composite structures.)

Moment Frame | A Moment Frame is a structural system in steel construction designed to resist lateral forces, such as those from earthquakes or wind. For example, in the construction of a high-rise building in a seismic zone, a steel moment frame provides stability and ensures the structure can withstand lateral loads. (Cross-reference with Flange for considerations related to the role of flanges in the stability of moment frames.)

Purlin | A Purlin is a horizontal structural member in a steel roof framing system that supports roof covering and transfers loads to the main structural frame. For instance, in the construction of a steel-framed warehouse, purlins are installed between rafters to provide support for the roofing material. (See Truss for insights into the use of purlins in truss systems.)

Riveted Connection | A Riveted Connection involves joining steel members using rivets. While less common in modern construction, riveted connections played a significant role in historic steel structures like bridges. For example, in the restoration of a historic bridge, riveted connections may be preserved or replaced with more modern methods like welding. (Refer to Bolted Connection for comparisons between riveted and bolted joining methods.)

Shear Connection | A Shear Connection is a type of connection in steel construction that allows the transfer of shear forces between connected members. For instance, in the design of a steel moment-resisting frame, shear connections enable the transfer of lateral forces and contribute to the overall stability of the structure. (See Cold-Formed Steel for considerations related to connecting cold-formed steel members.)

Stability | Stability in steel construction refers to the ability of a structure to maintain equilibrium and resist deformation or collapse under various loads. For example, in the design of a tall steel telecommunications tower, stability considerations are crucial to ensure the tower can withstand wind loads without excessive deflection. (Cross-reference with Tension Member for insights into stability considerations in tension members.)

Steel Frame | A Steel Frame is a structural system where steel members, such as columns and beams, are used to support building loads. For example, in the construction of a modern office building, a steel frame provides the primary structural support for the floors and roof. (See Built-Up Section for considerations related to creating steel members with enhanced load-carrying capabilities.)

Tension Member | A Tension Member in steel construction is a structural element subjected to axial tension forces. For instance, in the design of a suspension bridge, steel cables serve as tension members that support the bridge deck. (Refer to Stability for considerations related to ensuring stability in structures with tension members.)

Truss | A Truss is a structural framework composed of

interconnected elements, such as beams and triangles, designed to support loads over spans. For example, in the construction of a steel truss bridge, the truss structure efficiently distributes loads and provides strength. (Cross-reference with Purlin for insights into the role of purlins in supporting roof loads in truss systems.)

Welded Connection | A Welded Connection involves joining steel members using welding processes. It offers increased strength and is commonly used in modern steel construction. For instance, in the fabrication of a steel pipe rack for an industrial facility, welded connections provide robust joints. (See Bolted Connection for comparisons between welded and bolted joining methods.)

Wind Bracing | Wind Bracing in steel construction refers to elements, such as braces or walls, designed to resist lateral forces generated by wind. For example, in the design of a steel-framed high-rise building, wind bracing elements contribute to the overall stability of the structure, preventing excessive sway during wind events.

CHAPTER 10: MASONRY STRUCTURES AND CONSTRUCTION

Anchor Bolt | An Anchor Bolt is a threaded bolt embedded in masonry structures to provide a connection point for external elements, such as steel structures. For example, in the construction of a masonry building with a steel frame, anchor bolts are strategically placed in the masonry walls to secure the steel columns, ensuring stability. (Cross-reference with Reinforced Masonry for considerations related to reinforcing masonry with embedded elements.)

Bed Joint | The Bed Joint is the horizontal layer of mortar upon which a masonry unit is laid. In the construction of a brick wall, each brick is typically laid on a bed joint of mortar, creating a stable and cohesive structure. (Refer to Stack Bond for insights into the arrangement of masonry units that impacts the appearance and structural integrity of bed joints.)

Bond Beam | A Bond Beam is a horizontal structural element in masonry walls designed to distribute loads and reinforce the wall. For instance, in the construction of a reinforced masonry wall, bond beams are often placed at specific heights to enhance structural integrity. (See Reinforced Masonry for considerations related to reinforcing masonry elements, including bond beams.)

Cavity Wall | A Cavity Wall consists of two separate wythes (layers) of masonry with a gap or cavity between them. The gap provides insulation and prevents water penetration. For example, in the construction of an energy-efficient building, a cavity wall with appropriate insulation contributes to thermal efficiency. (Cross-reference with Weep Hole for considerations related to drainage in cavity walls.)

Collar Joint | A Collar Joint is a mortar joint between two masonry units that are vertically aligned. In the construction of a brick wall, collar joints help create a visually appealing and structurally sound pattern by aligning the vertical edges of adjacent bricks. (Refer to Stack Bond for insights into the arrangement of masonry units that affects the appearance and functionality of collar joints.)

Concrete Masonry Unit (CMU) | A Concrete Masonry Unit, commonly known as a CMU or concrete block, is a precast concrete block used in masonry construction. For example, in the construction of a residential building, concrete masonry units may be used to build exterior walls, providing both structural support and insulation.

Coping | Coping is the protective cap or covering on the top of a masonry wall to prevent water penetration and enhance durability. In the construction of a brick parapet wall, coping stones are often installed to shield the wall from weather elements and prevent water damage. (See Weep Hole for considerations related to drainage in masonry walls with coping.)

Course | A Course is a horizontal row of masonry units laid in a specific pattern to form a layer in a wall. In the construction of a brick wall, courses are arranged in a sequence to create a cohesive and visually appealing structure. (Refer to Stack Bond for insights into the arrangement of masonry units that affects the appearance and strength of courses.)

Lintel | A Lintel is a horizontal structural element, often made of steel or reinforced concrete, placed above openings, such as doors or windows, to support the load from the masonry above. For example, in the construction of a masonry facade with large windows, lintels provide essential structural support. (Cross-reference with Reinforced Masonry for considerations related to reinforcing lintels in masonry structures.)

Mortar | Mortar is a mixture of cement, sand, and water used to bond masonry units together. In the construction of a brick wall, mortar is applied between bricks to create a strong and stable structure. (See Bed Joint for insights into the horizontal layer of mortar that supports each masonry unit.)

Parge Coat | A Parge Coat is a thin coat of mortar or cement applied to the surface of a masonry wall for protective or decorative purposes. For instance, in the construction of a concrete block retaining wall, a parge coat may be applied to enhance the wall's appearance and protect it from the elements.

Pilaster | A Pilaster is a vertical column-like element incorporated into a masonry wall for structural support or decorative purposes. In the construction of a classical-style building, pilasters may be used to create a visually appealing facade and enhance the overall architectural design. (Refer to Column for considerations related to vertical load-bearing elements in masonry structures.)

Reinforced Masonry | Reinforced Masonry involves incorporating materials like steel reinforcement into masonry elements to enhance strength and durability. For example, in the construction of a masonry shear wall in a seismic zone, reinforced masonry provides added resistance against lateral forces. (Cross-reference with Lintel for insights into reinforcing lintels to support loads in masonry structures.)

Sill | A Sill is the horizontal member at the bottom of a window or door opening in a masonry wall. In the construction of a brick facade, sills provide support for windows and doors, and they often have a sloped surface to shed water away from the wall. (See Lintel for considerations related to horizontal structural elements above openings.)

Stack Bond | Stack Bond is a masonry unit arrangement where all vertical joints align, creating a symmetrical and uniform appearance. In the construction of a modern brick wall with a contemporary aesthetic, stack bond patterns are often employed to achieve a sleek and clean look. (Refer to Collar Joint for insights into the arrangement of masonry units that impacts the appearance and structure of stack bonds.)

Story Pole | A Story Pole is a marked rod used to measure and maintain consistent vertical spacing between courses of masonry units during construction. In the construction of a brick wall, a story pole helps ensure even courses and a visually pleasing outcome. (See Course for insights into horizontal rows of masonry units laid in a specific pattern.)

Tie Beam | A Tie Beam is a horizontal structural element that connects and stabilizes two masonry walls. For example, in the construction of a masonry building with multiple wings, tie beams may be used to provide lateral stability and prevent the walls from spreading apart. (Cross-reference with Collar Joint for considerations related to the connection of tie beams in masonry structures.)

Veneer | A Veneer is a thin layer of masonry, such as brick or stone, applied to the exterior of a building for decorative or protective purposes. In the construction of a modern commercial building, a brick veneer may be used to enhance the aesthetic appeal without providing structural support. (See Cavity Wall for considerations related to the construction of masonry veneer walls.)

Weep Hole | A Weep Hole is a small opening in a masonry wall that allows water to drain out, preventing water buildup and potential damage. For instance, in the construction of a brick facade with cavity walls, weep holes are strategically placed to ensure proper drainage and prevent moisture-related issues. (Cross-reference with Cavity Wall for insights into the drainage features of cavity walls.)

Wythe | A Wythe is a single vertical layer of masonry units in a wall. In the construction of a brick wall, multiple wythes are stacked horizontally to create a sturdy and visually appealing structure. (Refer to Cavity Wall for considerations related to the arrangement of multiple wythes with a gap in between.)

CHAPTER 11: TIMBER CONSTRUCTION AND APPLICATIONS

Bark | Bark is the protective outer layer of a tree trunk. In timber construction, the removal of bark is essential in the processing of raw timber into usable wood. For example, when preparing logs for lumber, the bark is typically stripped to expose the raw timber, which can then undergo further processing. (Cross-reference with Green Timber for considerations related to timber harvested with its bark intact.)

Beams | Beams are horizontal structural members that support loads in a building or structure. In timber construction, beams are often created by connecting timber elements horizontally to resist bending and provide structural support. For instance, in the construction of a timber-framed house, timber beams may support the weight of the floors and roof. (See Joists for insights into the role of beams in supporting other structural elements.)

Birdsmouth Joint | A Birdsmouth Joint, also known as a heel joint, is a notch cut into the base of a timber rafter to allow it to sit securely on a supporting beam. For example, in the construction of a timber roof, birdsmouth joints ensure a stable connection between rafters and supporting beams, preventing slippage. (Cross-reference with Joists for considerations related to the connection of joists with supporting beams.)

Bolts | Bolts are fasteners used in timber construction to join timber elements together securely. In the assembly of a timber truss, bolts are commonly employed to connect truss members and ensure structural integrity. For instance, bolts may be used to connect the bottom chord of a truss to a supporting beam. (See Trusses for insights into the use of bolts in the assembly of timber trusses.)

Box Frame | A Box Frame is a type of timber frame construction where the structural members form a rectangular box-like shape. In the construction of a timber box frame house, this design provides stability and allows for open floor plans. For example, a box frame may consist of timber posts at the corners and horizontal beams connecting them. (See Post-and-Beam for considerations related to the use of vertical and horizontal timber members.)

Cantilever | A Cantilever is a structural element that projects horizontally beyond its support. In timber construction, a cantilevered timber beam or joist may extend beyond a supporting post or wall. For instance, in the construction of a timber balcony, a cantilevered beam provides support for the extended platform without the need for additional supports. (See Beams for insights into the role of beams in supporting loads.)

Cross-Laminated Timber (CLT) | Cross-Laminated Timber (CLT) is a type of engineered wood product composed of layers of timber boards glued together in a crosswise pattern. In the construction of a CLT building, large panels of CLT are used as structural elements, providing strength and stability. For example, CLT panels may serve as the floors and walls of a multi-story timber building. (See Laminated Veneer Lumber (LVL) for insights into another type of engineered wood product.)

Dovetail Joint | A Dovetail Joint is a type of woodworking joint that interlocks with a series of wedge-shaped projections. In timber construction, dovetail joints are often used to connect timber elements, such as the corners of timber frames. For instance, in the construction of a timber chest, dovetail joints provide strength and visual appeal to the corners. (Cross-reference with Mortise and Tenon Joint for considerations related to another type of woodworking joint.)

Flitch Beam | A Flitch Beam is a composite beam made by sandwiching a steel plate between two timber beams. In timber construction, flitch beams combine the strength of steel with the aesthetic appeal of timber. For example, in the construction of a timber-framed bridge, flitch beams may be used to span longer distances with increased load-bearing capacity. (See Steel Members for insights into the use of steel in combination with timber.)

Grain | Grain in timber refers to the direction of the fibers in the wood. Understanding the grain is crucial in timber construction, as it influences the strength and behavior of the wood. For example, in the construction of a timber beam, the grain orientation is considered to ensure optimal load-bearing capacity. (Cross-reference with Warp for considerations related to the deformation of timber due to grain characteristics.)

Green Timber | Green Timber refers to freshly harvested timber that has not undergone the drying or seasoning process. In timber construction, the use of green timber may lead to shrinkage and warping as the wood dries. For instance, in the construction of a timber-framed structure, allowing green timber to acclimate before use helps prevent subsequent deformation. (See Bark for considerations related to timber harvested with its bark intact.)

Heartwood | Heartwood is the inner, non-living portion of a tree trunk that often has a darker color. In timber construction, heartwood is preferred for its increased density and durability. For example, in the construction of exterior timber elements like posts or beams, selecting heartwood helps enhance resistance to decay and insect infestation. (See Sapwood for considerations related to the outer, living portion of a tree trunk.)

Joists | Joists are horizontal structural members that support the floor or ceiling loads. In timber construction, joists are commonly used to create a framework for flooring. For example, in the construction of a timber-framed house, joists may support the flooring system, providing a stable surface. (See Beams for insights into the role of beams in supporting other structural elements.)

Laminated Veneer Lumber (LVL) | Laminated Veneer Lumber (LVL) is an engineered wood product made by bonding together thin veneers of wood with adhesive. In timber construction, LVL is used for its enhanced strength and versatility. For example, in the construction of a timber beam, LVL may be used to achieve greater load-bearing capacity. (See Cross-Laminated Timber (CLT) for insights into another type of engineered wood product.)

Live Load | Live Load refers to the temporary and variable loads that a structure may experience, such as people, furniture, or snow. In timber construction, accounting for live loads is crucial in ensuring structural stability. For instance, in the construction of a timber floor, the design considers the expected live loads to determine the required load-bearing capacity. (See Dead Load for insights into permanent and stationary loads.)

Mortise and Tenon Joint | A Mortise and Tenon Joint is a traditional woodworking joint where a projection (tenon) on one timber piece fits into a corresponding hole (mortise) in another. In timber construction, mortise and tenon joints are often used for strong and durable connections. For example, in the construction of a timber door frame, mortise and tenon joints provide stability and longevity. (Cross-reference with

Dovetail Joint for considerations related to another type of woodworking joint.)

Plywood | Plywood is a sheet material composed of thin layers (plies) of wood veneer glued together. In timber construction, plywood is used for various applications, including sheathing and structural support. For example, in the construction of a timber-framed wall, plywood sheathing provides additional strength and stability. (See Veneer for insights into the use of thin layers of wood in construction.)

Post-and-Beam | Post-and-Beam construction involves vertical posts supporting horizontal beams, creating an open and airy structure. In the construction of a timber post-and-beam house, this design allows for large windows and open spaces. For example, post-and-beam construction may feature large timber posts supporting beams and creating a visually appealing interior. (See Box Frame for considerations related to another type of timber frame construction.)

Ring Beam | A Ring Beam is a horizontal structural element that encircles a structure, often tying together vertical posts or walls. In timber construction, a ring beam may be used to provide lateral stability. For instance, in the construction of a timber roundhouse, a ring beam may connect the tops of supporting posts, preventing lateral movement. (See Post-and-Beam for insights into the use of vertical posts in combination with horizontal beams.)

Sapwood | Sapwood is the outer, living portion of a tree trunk that often has a lighter color. In timber construction, sapwood is generally less dense and durable than heartwood. For example, in the construction of interior timber elements like trim, using sapwood may be acceptable due to lower exposure to decay and insects. (See Heartwood for considerations related to the inner, non-living portion of a tree trunk.)

Shear Wall | A Shear Wall is a vertical structural element that resists lateral forces, such as wind or seismic loads. In timber construction, shear walls may be constructed using plywood or other materials to provide stability. For example, in the construction of a timber-framed house, shear walls may be strategically placed to resist horizontal forces. (See Warp for considerations related to the potential deformation of timber under lateral loads.)

Shingles | Shingles are small, thin pieces of wood used to cover roofs or walls in timber construction. For example, in the construction of a timber-clad house, shingles may be applied to the exterior for weather protection and aesthetic appeal. (See Cladding for insights into the use of materials to cover the exterior of a timber structure.)

Stave Construction | Stave Construction involves assembling vertical staves or planks to create cylindrical structures, such as tanks or silos. In timber construction, stave construction may be employed for unique architectural elements. For instance, in the construction of a timber water tower, stave construction allows for the creation of a cylindrical container. (See Ring Beam for considerations related to horizontal elements in cylindrical structures.)

Tension Member | A Tension Member is a structural element subjected to tension forces. In timber construction, tension members may include cables or ropes used to support loads. For example, in the construction of a timber bridge, tension members may be employed to suspend the bridge deck. (Cross-reference with Compression Member for considerations related to structural elements subjected to compression forces.)

Tongue and Groove | Tongue and Groove is a type of joint where a protruding "tongue" on one timber piece fits into a groove on another. In timber construction, tongue and groove joints are often used in flooring or paneling. For example, in the construction of a timber floor, tongue and groove boards interlock to create a stable and smooth surface. (See Flooring for insights into the use of timber in creating functional and aesthetic flooring.)

Trusses | Trusses are structural frameworks made of triangular units connected at joints. In timber construction, trusses are often used to support roof loads. For example, in the construction of a timber-framed barn, trusses may be employed to create a sturdy and efficient roof structure. (See Bolts for considerations related to the use of fasteners in connecting truss members.)

Veneer | Veneer is a thin layer of wood sliced or peeled from a log and used to cover surfaces. In timber construction, veneer may be applied to achieve a desired appearance. For example, in the construction of a timber door, veneer may be used to enhance the visual appeal of the door. (See Plywood for insights into the use of veneer in creating sheet material.)

Warp | Warp refers to the deformation of timber caused by uneven drying or changes in moisture content. In timber construction, preventing warp is essential for maintaining structural integrity. For example, in the construction of a timber floor, proper acclimation of the timber and careful installation help minimize the risk of warp. (See Grain for considerations related to the direction of fibers in timber.)

Wood Preservatives | Wood Preservatives are substances applied to timber to protect it from decay, insects, and fungi. In timber construction, the use of preservatives extends the lifespan of timber elements. For example, in the construction of outdoor timber structures like decks, applying wood preservatives helps prevent deterioration due to exposure to the elements. (See Green Timber for considerations related to freshly harvested timber.)

CHAPTER 12: EARTHWORK AND EXCAVATION TECHNIQUES

Backfilling | Backfilling is the process of refilling an excavated area with earth or other suitable materials. In construction, backfilling is commonly done around foundations to provide support and improve soil stability. For example, after laying a utility pipe in a trench, backfilling is performed to restore the excavated area to its original condition. (Cross-reference with Excavation for insights into the initial process of digging or removing earth.)

Benching | Benching involves creating a series of steps or horizontal levels in an excavation to control slope stability. This technique is commonly used in open-pit mining and excavation projects. For instance, when excavating a hillside for construction, benching helps prevent soil erosion and slope failure. (Cross-reference with Slope Stability for considerations related to maintaining the stability of excavated slopes.)

Borrow Pit | A Borrow Pit is an excavation site where soil, gravel, or other materials are extracted for use elsewhere in a construction project. For example, in road construction, a borrow pit may be established to obtain suitable materials for road embankments. (Cross-reference with Cut and Fill for insights into the process of excavation and redistribution of soil within a construction site.)

Compaction | Compaction is the process of mechanically reducing the volume of soil through the application of pressure. This technique improves soil density and stability. In road construction, compaction is crucial for ensuring a solid foundation. For instance, a compactor may be used to compress soil layers during the construction of a highway to prevent settlement. (Cross-reference with Soil Compaction for considerations related to the specific techniques used to compact soil.)

Culvert | A Culvert is a structure that allows water to flow beneath a road, railway, or embankment. Culverts are essential for managing water drainage in construction projects. For example, during the construction of a road, culverts are installed to facilitate the passage of water under the road, preventing flooding and erosion. (Cross-reference with Water Diversion for insights into techniques used to control and redirect water in construction.)

Cut and Fill | Cut and Fill is a construction technique involving the excavation of soil from one area (cut) and using it to fill another area (fill). This method is often employed to create level surfaces for construction projects. For instance, when building a residential development on uneven terrain, cut and fill operations are performed to achieve a uniform ground level. (Cross-reference with Borrow Pit for considerations related to obtaining materials for filling excavated areas.)

Earthmoving Equipment | Earthmoving Equipment includes a variety of heavy machinery used for excavation, grading, and other earthmoving tasks in construction. Examples of earthmoving equipment include excavators, bulldozers, and graders. These machines are crucial for efficiently moving large volumes of soil during construction projects. (Cross-reference with Excavation for insights into the initial process of digging or removing earth.)

Excavation | Excavation is the process of digging, trenching, or otherwise removing earth to create a void or cavity in the ground. Excavation is a fundamental step in many construction projects. For example, when preparing a construction site for a building foundation, excavation is done to reach the required depth. (Cross-reference with Backfilling for considerations related to refilling excavated areas.)

Foundation Excavation | Foundation Excavation specifically refers to the excavation work carried out to prepare the ground for the installation of a building's foundation. This process involves digging to the prescribed depth and dimensions to accommodate the foundation design. For instance, in residential construction, foundation excavation is done to create space for footings and other foundation components. (Cross-reference with Compaction for insights into preparing a stable foundation through soil compaction.)

Grading | Grading is the process of adjusting the slope and contour of the land to achieve a desired level or profile. Grading is commonly done for landscaping, road construction, or preparing building sites. For example, in the construction of a parking lot, grading is performed to ensure proper drainage and a level surface. (Cross-reference with Benching for considerations related to creating stepped or terraced slopes.)

Land Clearing | Land Clearing involves the removal of vegetation, trees, rocks, and other obstacles from a construction site. This preparatory step is essential for creating a clean and usable space for construction activities. For instance, before building a residential community, land clearing is conducted to remove trees and vegetation from the designated area. (Cross-reference with Cut and Fill for insights into the redistribution of cleared materials within the construction site.)

Lift Thickness | Lift Thickness refers to the depth or thickness of soil or material compacted in a single layer during the compaction process. The lift thickness is a critical factor in achieving the desired soil density. For example, in road construction, engineers specify the lift thickness to ensure proper compaction and stability of the roadbed. (Cross-reference with Compaction for considerations related to achieving optimal soil density.)

Overexcavation | Overexcavation occurs when the excavation process extends beyond the specified dimensions or depth. This can lead to additional costs and challenges in construction projects. For instance, in the construction of a basement, overexcavation may occur if the excavation team digs deeper than the intended foundation depth. (Cross-reference with Foundation Excavation for insights into the specific excavation work related to foundation preparation.)

Slope Stability | Slope Stability refers to the ability of an excavated slope or embankment to maintain its structural integrity and resist sliding or collapsing. Ensuring slope stability is crucial to prevent landslides or erosion. For example, in highway construction on hilly terrain, engineers implement measures to enhance slope stability and prevent soil movement. (Cross-reference with Benching for considerations related to creating stepped slopes to improve stability.)

Stripping | Stripping is the process of removing the top layer of soil or vegetation from a construction site. This is often done to expose the underlying soil for further construction activities. For instance, before building a foundation, stripping is performed to clear the site and expose the natural ground. (Cross-reference with Land Clearing for insights into the broader process of preparing a construction site by removing obstacles and vegetation.)

Subgrade | Subgrade refers to the prepared natural ground or surface on which a construction project is built. Achieving proper subgrade compaction is essential for the stability of the entire structure. For example, in road construction, engineers focus on preparing a stable subgrade to support the road pavement and ensure longevity. (Cross-reference with Compaction for considerations related to achieving optimal subgrade compaction.)

Swale | A Swale is a shallow, vegetated channel designed to manage and direct stormwater runoff. Swales are often used in landscaping and construction projects to control water drainage. For instance, in the construction of a residential community, swales may be incorporated to channel rainwater away from buildings and prevent flooding. (Cross-reference with Culvert for insights into structures used to manage water flow beneath roads or embankments.)

Trenching | Trenching is the process of digging long, narrow excavations in the ground, typically for installing utilities or foundations. Trenches are commonly used in construction to accommodate pipelines, cables, or footings. For example, in the construction of a sewer line, trenching is performed to create a pathway for laying the pipes. (Cross-reference with Excavation for insights into the broader process of digging or removing earth.)

Undercutting | Undercutting is the removal of soil beneath the level of the existing ground surface. This technique is used to create a stable foundation or improve slope stability. For example, in the construction of a retaining wall, undercutting may be performed to provide additional support and prevent the wall from collapsing. (Cross-reference with Slope Stability for considerations related to enhancing stability through excavation techniques.)

Vibro Compaction | Vibro Compaction is a ground improvement technique that involves the use of vibrating probes to densify loose or granular soils. This process increases soil density and stability. For example, in the construction of a high-rise building, vibro compaction may be employed to enhance the bearing capacity of the soil beneath the foundation. (Cross-reference with Compaction for insights into the broader process of mechanically reducing soil volume.)

Water Diversion | Water Diversion involves directing or redirecting water away from a construction site to prevent water-related issues such as flooding or erosion. Various techniques, including the use of swales or culverts, may be employed for effective water diversion. For example, during the construction of a bridge, water diversion measures are implemented to channel water away from the construction area. (Cross-reference with Culvert for insights into structures designed to manage water flow beneath roads or embankments.)

Xeriscaping | Xeriscaping is a landscaping method that focuses on creating water-efficient and drought-resistant outdoor spaces. This approach is particularly relevant in arid regions. For instance, in the construction of a commercial complex in a desert area, xeriscaping may be employed to design outdoor spaces that require minimal water usage. (Cross-reference with Water Diversion for considerations related to managing water resources in landscaping.)

Yield Stress | Yield Stress is the point at which a material undergoes significant deformation or begins to flow. In earthwork and excavation, understanding the yield stress of soils is crucial for determining their behavior under load. For example, in the construction of a dam, engineers assess the yield stress of foundation soils to ensure stability against the imposed load. (Cross-reference with Compaction for insights into achieving optimal soil density and yield stress.)

Zoning | Zoning refers to the division of land into different zones or designated areas based on specific land use regulations. Zoning is a critical aspect of urban planning and construction. For example, in the construction of a mixed-use development, zoning regulations dictate where residential, commercial, or industrial structures can be located. (Cross-reference with Land Clearing for considerations related to preparing a construction site within specified zoning regulations.)

CHAPTER 13: CONSTRUCTION METHODS AND EQUIPMENT

Assembly Line Construction | Assembly Line Construction refers to a systematic approach where construction tasks are divided into specialized, sequential operations, mirroring the efficiency of manufacturing assembly lines. This method enhances productivity by streamlining processes. For instance, in the construction of prefabricated homes, assembly line techniques are employed to efficiently produce standardized components, reducing construction time. (Cross-reference with Offsite Construction for insights into the broader use of prefabrication and assembly in construction.)

BIM (Building Information Modeling) | Building Information Modeling (BIM) is a digital representation of the physical and functional characteristics of a construction project. It integrates various aspects, including design, construction, and operation. For example, during the construction of a commercial building, BIM facilitates collaborative planning, helping teams visualize and analyze project elements in a shared digital environment. (Cross-reference with Green Building for considerations related to sustainability in construction.)

Caisson | A Caisson is a watertight structure used in construction for underwater foundations, especially in bridge piers or other structures requiring deep foundations. For instance, in the construction of a bridge over a river, caissons may be employed to create a stable foundation underwater, allowing construction to proceed in challenging conditions. (Cross-reference with Foundation for insights into various foundation types used in construction.)

Crane | A Crane is a lifting machine equipped with cables and pulleys, used to move heavy materials vertically and horizontally on construction sites. In the construction of high-rise buildings, cranes play a vital role in lifting and placing construction materials, such as steel beams and concrete panels, at different elevations. (Cross-reference with Heavy Equipment for considerations related to various types of heavy machinery used in construction.)

Demolition | Demolition involves the systematic dismantling or destruction of structures and buildings. For example, in urban redevelopment projects, demolition is carried out to clear old structures, making space for new construction. Demolition methods vary, ranging from manual dismantling to the use of specialized equipment like wrecking balls. (Cross-reference with Construction Methods for insights into the broader techniques used in construction processes.)

Elevator (Construction Hoist) | In construction, an Elevator, or Construction Hoist, is a vertical transportation system used to move workers and materials between different levels of a construction site. For instance, during the construction of a skyscraper, construction hoists are employed to transport workers and construction materials efficiently, ensuring timely completion. (Cross-reference with Infrastructure for considerations related to essential systems supporting construction operations.)

Formwork | Formwork is the temporary mold or structure used to support and shape concrete until it hardens. For example, when constructing a reinforced concrete foundation, formwork is used to create the desired shape. The formwork is later removed, leaving behind the hardened concrete structure. (Cross-reference with Reinforcement for insights into the use of materials to strengthen and support structures in construction.)

Green Building | Green Building focuses on sustainable construction practices that minimize environmental impact. For instance, in the construction of a LEED-certified office building, green building principles are applied, incorporating energy-efficient systems, renewable materials, and environmentally friendly construction methods. (Cross-reference with BIM for considerations related to digital tools supporting sustainable design and construction.)

Heavy Equipment | Heavy Equipment encompasses a variety of large machinery used in construction, including excavators, bulldozers, and cranes. In the construction of a highway, heavy equipment is utilized for tasks such as earthmoving, grading, and lifting heavy materials, contributing to the efficiency and speed of the construction process. (Cross-reference with Crane for insights into specific machinery used for lifting and moving materials.)

Infrastructure | Infrastructure refers to the essential systems and facilities supporting construction operations, such as roads, utilities, and communication networks. For example, in the construction of a residential development, infrastructure planning involves designing roads and utilities to serve the community. (Cross-reference with Elevator (Construction Hoist) for considerations related to vertical transportation systems supporting construction activities.)

Jack and Bore | Jack and Bore is a trenchless construction method used to install pipes or conduits underground without extensive excavation. For instance, when installing utility lines beneath a busy city street, jack and bore techniques are employed to minimize disruption to traffic and reduce the need for extensive excavation. (Cross-reference with Tunnel Boring Machine (TBM) for insights into other trenchless methods used in construction.)

Kicker | A Kicker, in construction, refers to a short, vertical wall section used to form a right-angle intersection between a vertical surface and a horizontal slab. For example, in the construction of a parking garage, kickers are utilized to create the junction between the vertical columns and the horizontal floor slabs. (Cross-reference with Formwork for considerations related to temporary molds used in concrete construction.)

Lift Slab Construction | Lift Slab Construction is a method where entire building slabs are cast on the ground and then lifted into position. For example, in the construction of a large warehouse, lift slab techniques may be employed to expedite the construction process by assembling the building's floors on the ground and then raising them into place. (Cross-reference with Assembly Line Construction for insights into systematic approaches to construction.)

Modular Construction | Modular Construction involves building sections or modules offsite and assembling them on the construction site. In the construction of a hotel, modular construction methods are utilized to prefabricate rooms offsite, allowing for quicker on-site assembly and reducing construction time. (Cross-reference with Assembly Line Construction for considerations related to systematic construction approaches.)

Nanotechnology in Construction | Nanotechnology in Construction involves the use of nanomaterials and techniques to enhance construction materials' properties. For example, in the construction of a high-performance concrete structure, nanotechnology may be applied to improve the material's strength and durability at the molecular level. (Cross-reference with Construction Methods for insights into the broader techniques used in construction processes.)

Offsite Construction | Offsite Construction refers to the prefabrication of building components away from the construction site. For instance, in the construction of a residential development, offsite construction may involve fabricating wall panels in a factory and transporting them to the site for assembly. (Cross-reference with Modular Construction for considerations related to assembling prefabricated modules.)

Pile Driving | Pile Driving is a method of installing deep foundation elements, called piles, into the ground to provide support for structures. For example, in the construction of a bridge over water, pile driving is employed to secure foundation piles into the riverbed, ensuring stability. (Cross-reference with Caisson for insights into specialized foundations used in challenging conditions.)

Quoin | In construction, a Quoin is an external corner or edge of a building, often accentuated or differentiated from the rest of the structure. For instance, in the construction of a historic building, quoins may be used to add architectural detail to the corners, enhancing the building's visual appeal. (Cross-reference with Architecture for considerations related to the design and aesthetics of structures.)

Reinforcement | Reinforcement involves adding materials, such as steel bars, to concrete to improve its strength and durability. For example, in the construction of a high-rise building, reinforcement is crucial to provide structural integrity and resist forces like wind and seismic activity. (Cross-reference with Formwork for insights into temporary molds used in concrete construction.)

Scaffolding | Scaffolding is a temporary structure used to support workers and materials during construction, maintenance, or repair activities. For instance, in the construction of a commercial building, scaffolding is erected to provide access to different levels, facilitating tasks such as painting and facade installation. (Cross-reference with Construction Methods for considerations related to various techniques used in construction processes.)

Tunnel Boring Machine (TBM) | A Tunnel Boring Machine (TBM) is a mechanized excavation tool used to create tunnels. For example, in the construction of an underground subway system, TBMs are employed to excavate tunnels efficiently. (Cross-reference with Jack and Bore for insights into trenchless methods used in construction.)

Underpinning | Underpinning is the process of strengthening and stabilizing the foundation of an existing structure. For instance, in the construction of a renovated historic building, underpinning may be necessary to ensure the foundation can support new loads and modifications. (Cross-reference with Foundation for considerations related to various foundation types used in construction.)

Value Stream Mapping | Value Stream Mapping is a lean management technique used to visualize and analyze the steps involved in delivering value during construction processes. For example, in the construction of a manufacturing facility, value stream mapping helps identify and eliminate non-value-added activities, improving overall efficiency. (Cross-reference with Lean Construction for insights into minimizing waste and optimizing efficiency in construction.)

Water Jetting | Water Jetting is a construction technique that uses high-pressure water jets for tasks such as surface cleaning, concrete removal, or cutting. For instance, in the construction of a bridge, water jetting may be employed to remove old paint or prepare surfaces for repairs. (Cross-reference with Construction Methods for considerations related to various techniques used in construction processes.)

XRF (X-ray Fluorescence) in Construction | X-ray Fluorescence (XRF) in Construction involves using XRF technology to analyze the composition of construction materials. For example, in the construction of a bridge, XRF analysis may be applied to assess the quality and composition of steel components. (Cross-reference with Quality Control for insights into ensuring construction materials meet specified standards.)

Yard | In construction, a Yard refers to an area where construction materials, equipment, or vehicles are stored. For instance, in the construction of a residential development, a construction yard may be established to organize and store building materials, ensuring efficient access for construction activities. (Cross-reference with Logistics for considerations related to the organized movement and storage of materials.)

Zero Energy Building | A Zero Energy Building is designed to produce as much energy as it consumes, resulting in a net-zero energy impact. For example, in the construction of an eco-friendly office building, zero energy building principles may be applied, incorporating renewable energy sources and energy-efficient systems. (Cross-reference with Sustainable Construction for considerations related to environmentally conscious construction practices.)

PART III: BUILDING CONSTRUCTION SYSTEMS AND COMPONENTS

CHAPTER 14: FOUNDATION SYSTEMS AND DESIGN

Anchor Bolt | Anchor bolts are threaded rods used to securely attach structures to concrete foundations. For instance, in the construction of a high-rise building, anchor bolts play a crucial role in anchoring the steel or concrete structure to the foundation, providing stability against lateral forces. (Cross-reference with Structural Engineering for insights into the broader principles of structural design.)

Bearing Capacity | Bearing capacity refers to the ability of the soil to support the loads imposed by a structure. In foundation design, understanding the bearing capacity is essential to ensure that the soil can withstand the loads without excessive settlement. For example, in the construction of a residential home, determining the bearing capacity of the soil helps in selecting an appropriate foundation type. (Cross-reference with Geotechnical Engineering for considerations related to soil mechanics.)

Cantilever Foundation | A cantilever foundation is a type of footing that projects beyond the supported column or wall. In the context of a bridge construction, a cantilever foundation may be employed to support the ends of bridge segments, allowing for the construction of extended spans.

(Cross-reference with Bridge Engineering for insights into the specialized considerations in bridge construction.)

Deep Foundation | Deep foundations extend into the underlying soil or rock to support structures where shallow foundations may be inadequate. An example is the use of pile foundations in the construction of a waterfront structure where the soil near the surface may not provide sufficient support. (Cross-reference with Pile Foundation for considerations related to specific deep foundation techniques.)

Earthquake Resistant Foundation | An earthquake-resistant foundation is designed to withstand the dynamic forces generated by seismic activity. In the construction of a seismic-resistant building, the foundation may incorporate features such as base isolators or flexible elements to absorb and dissipate earthquake-induced forces. (Cross-reference with Seismic Design for insights into broader strategies for earthquake-resistant construction.)

Footing | A footing is a structural component that distributes the load from a column or wall to the underlying soil. In a residential construction project, footings are essential for supporting the weight of the house and transmitting it to the ground, preventing settlement. (Cross-reference with Structural Design for considerations related to the overall load path in structures.)

Geotechnical Engineering | Geotechnical engineering involves the study of soil and rock mechanics to understand their behavior under various conditions. In the construction of a transportation infrastructure project, geotechnical engineering helps assess soil properties to ensure the stability of foundations and slopes. (Cross-reference with Transportation Engineering for insights into the unique challenges of infrastructure projects.)

Helical Pile | A helical pile is a deep foundation element consisting of a helical-shaped screw plate attached to a steel shaft. In the construction of a telecommunications tower, helical piles may be used to anchor the tower foundation securely, especially in areas with challenging soil conditions. (Cross-reference with Communication Tower Foundation Design for considerations specific to tower foundations.)

Isolated Footing | An isolated footing is a type of shallow foundation that supports a single column or load. For example, in the construction of an industrial facility, isolated footings may be used to support individual machine foundations, distributing loads to the soil. (Cross-reference with Machine Foundation Design for insights into the specialized considerations for machinery support.)

Jet Grouting | Jet grouting is a ground improvement technique where a high-pressure jet of grout is used to mix and strengthen the soil in situ. In the construction of a waterfront structure, jet grouting may be applied to improve soil stability and prevent erosion along the shoreline. (Cross-reference with Ground Improvement Techniques for a broader understanding of soil stabilization methods.)

Keller's Pressure Balance Method | Keller's Pressure Balance Method is a foundation construction technique that involves balancing the pressure of injected grout with the soil's resistance. In tunnel construction beneath an existing structure, Keller's Pressure Balance Method may be used to minimize ground settlement and maintain structural integrity. (Cross-reference with Tunneling Construction for insights into tunnel construction methods.)

Lateral Earth Pressure | Lateral earth pressure is the force exerted by soil against a retaining structure, such as a basement wall. In the construction of a residential development on a sloping site, understanding lateral earth pressure is crucial for designing retaining walls that can resist soil pressure and prevent slope instability. (Cross-reference with Retaining Wall Design for considerations related to retaining structures.)

Mat Foundation | A mat foundation, also known as a raft foundation, is a type of shallow foundation that covers the entire building footprint. In the construction of a large commercial complex, a mat foundation may be employed to evenly distribute the building loads and minimize differential settlement. (Cross-reference with Commercial Building Foundation Design for insights into considerations specific to large-scale structures.)

Neotectonics | Neotectonics involves the study of recent and active geological processes, providing insights into the present-day deformation of the Earth's crust. In the construction of a geological research facility, considerations of neotectonics may influence foundation design to account for potential ground movement. (Cross-reference with Geological Engineering for a broader understanding of geological influences on construction.)

Over-excavation | Over-excavation is the process of digging deeper than required for a foundation to remove weak or unsuitable soil. In the construction of a high-rise building, over-excavation may be performed to reach more stable soil layers, ensuring a secure foundation. (Cross-reference with Excavation Techniques for insights into various excavation methods.)

Pier Foundation | A pier foundation consists of cylindrical vertical columns that support beams or slabs. For example, in the construction of a boardwalk over wetlands, pier foundations may be used to elevate the structure and minimize environmental impact. (Cross-reference with Wetland Construction for considerations related to construction in environmentally sensitive areas.)

Quasi-Static Testing | Quasi-static testing involves applying slow and controlled loads to assess the behavior of foundations. In the construction of a bridge, quasi-static testing may be used to evaluate the performance of bridge foundations under gradual loading conditions, providing valuable data for design verification. (Cross-reference with Bridge Foundation Design for insights into considerations specific to bridge foundations.)

Raft Foundation | A raft foundation, also known as a mat foundation, is a type of shallow foundation that covers the entire building footprint. In the construction of a residential neighborhood, a raft foundation may be employed to provide a stable base for multiple homes with shared walls. (Cross-reference with Residential Building Foundation Design for considerations specific to housing developments.)

Soil Settlement | Soil settlement refers to the gradual sinking or compression of soil under a load. In the construction of a highway bridge, engineers must consider soil settlement to prevent uneven subsidence that could affect the bridge's structural integrity. (Cross-reference with Bridge Construction for insights into considerations specific to bridge projects.)

Tie Beam | A tie beam is a horizontal member that connects two or more columns to provide lateral stability. In the construction of an industrial facility, tie beams may be incorporated into the foundation design to resist horizontal forces and ensure overall stability. (Cross-reference with Industrial Facility Foundation Design for considerations specific to industrial structures.)

Underpinning | Underpinning is the process of strengthening and stabilizing the foundation of an existing structure. In the construction of a historic building undergoing renovation, underpinning may be necessary to address foundation issues and ensure the structure's longevity. (Cross-reference with Historic Building Restoration for insights into considerations specific to preserving and renovating historic structures.)

Vertical Drain | A vertical drain, also known as a wick drain, is a geotechnical solution used to accelerate soil consolidation. In the construction of an airport runway on soft soil, vertical drains may be installed to reduce settlement and improve soil stability. (Cross-reference with Airport Construction for insights into considerations specific to airport infrastructure.)

Waler Beam | A waler beam is a horizontal support element used to brace vertical retaining walls or excavation support systems. In the construction of a deep basement for a commercial building, waler beams may be employed to provide lateral support and prevent wall movement. (Cross-reference with Basement Construction for insights into considerations specific to constructing below-ground structures.)

Xerophyte | Xerophytes are plants adapted to survive in arid conditions. In the construction of a botanical garden in a desert region, understanding the presence of xerophytes may influence foundation design to minimize disruption to the natural ecosystem. (Cross-reference with Botanical Garden Construction for insights into considerations specific to creating sustainable and environmentally conscious landscapes.)

Yield Line Analysis | Yield line analysis is a structural engineering method used to assess the collapse mechanism in slabs and plates. In the construction of a parking garage, yield line analysis may be applied to evaluate the structural integrity of concrete slabs supporting vehicular loads. (Cross-reference with Parking Garage Design for insights into considerations specific to designing structures for vehicular use.)

Zero-Lot-Line House | A zero-lot-line house is constructed close to the property line, maximizing the use of available space. In urban residential construction, zero-lot-line houses may be built to optimize land use in densely populated areas, requiring careful consideration of foundation design to ensure stability without encroaching on neighboring properties.

(Cross-reference with Urban Construction for insights into considerations specific to construction in urban environments.)

CHAPTER 15: STRUCTURAL SYSTEMS AND DESIGN FOR BUILDINGS

Architectural Engineering | Architectural engineering is a field that integrates the principles of architecture and engineering to create structurally sound and aesthetically pleasing buildings. In the design of a modern office building, architectural engineering ensures that the structural elements align with the architectural vision, creating a harmonious and functional structure. (Cross-reference with Building Design for insights into the holistic design approach.)

Base Shear | Base shear is the total lateral force exerted at the base of a structure during an earthquake. In the seismic design of a high-rise residential tower, engineers consider base shear to dimension structural elements and ensure the building's stability under seismic forces. (Cross-reference with Seismic Design for considerations specific to earthquake-resistant design.)

Cantilever | A cantilever is a structural element supported at only one end, projecting horizontally into space. For instance, in the construction of a contemporary art museum, a cantilevered section may be incorporated to create an innovative and visually striking design while providing sheltered spaces below. (Cross-reference with Architectural Elements for insights into design features.)

Dead Load | Dead load refers to the static weight of a structure's permanent components, such as walls and floors. When designing a residential building, dead load considerations ensure that the structural system can support the weight of the building itself, along with fixed elements like interior partitions. (Cross-reference with Structural Design for insights into load calculations.)

Elasticity | Elasticity is the property of a material to return to its original shape after deformation. In the construction of a sports stadium roof, understanding the elasticity of materials like steel allows engineers to design flexible structures that can withstand dynamic loads without permanent deformation. (Cross-reference with Material Science for insights into the mechanical properties of construction materials.)

Flexural Strength | Flexural strength is the ability of a material to resist bending. When designing a pedestrian bridge, engineers consider the flexural strength of materials like reinforced concrete or steel to ensure the bridge can support the applied loads and remain structurally sound. (Cross-reference with Bridge Design for considerations specific to bridge structures.)

Gravity Load | Gravity load is the force exerted on a structure due to the weight of its components. In the construction of a residential building, architects and engineers account for gravity loads to design foundations and structural systems capable of supporting the structure's weight. (Cross-reference with Foundation Design for insights into load-bearing considerations.)

Hinge | In structural engineering, a hinge represents a

theoretical point of rotation allowing movement in response to applied forces. When designing a large industrial facility with overhead cranes, engineers may use hinged connections to accommodate movement and prevent structural damage. (Cross-reference with Industrial Facility Design for insights into considerations specific to industrial structures.)

Internal Forces | Internal forces are the stresses and deformations that occur within a structure due to applied loads. Understanding internal forces is crucial in the design of a high-rise office building, where engineers analyze how different structural elements respond to various loads to ensure structural integrity. (Cross-reference with Structural Analysis for insights into the analysis of internal forces.)

Joist | A joist is a horizontal structural member used to support the floor or ceiling of a building. In residential construction, joists are commonly employed to create a stable and level framework for the floors, ensuring the structural integrity of the entire building. (Cross-reference with Residential Construction for insights into considerations specific to housing.)

Knee Brace | A knee brace is a diagonal structural element used to resist lateral loads. When designing a school building, knee braces may be incorporated into the structural system to provide additional stability and prevent excessive lateral movement during seismic events. (Cross-reference with Educational Facility Design for insights into considerations specific to school buildings.)

Live Load | Live load refers to the dynamic forces exerted on a structure due to the presence of people, furniture, or movable equipment. In the design of a shopping mall, architects and engineers consider live loads to ensure that the structural system can safely support the varying weights and movements of occupants and merchandise. (Cross-reference with Commercial Building Design for insights into considerations specific to commercial structures.)

Moment Resisting Frame | A moment resisting frame is a structural system that resists lateral forces through the development of bending moments. In the construction of a hospital, moment resisting frames may be utilized to enhance the building's seismic performance, providing flexibility and stability. (Cross-reference with Hospital Design for insights into considerations specific to healthcare facilities.)

Nodal Point | A nodal point is a location in a structure where several members meet. In the design of a convention center with a large open space, engineers focus on nodal points to ensure the proper connection of structural elements, allowing for the efficient transfer of loads. (Cross-reference with Convention Center Design for insights into considerations specific to large event venues.)

Orthotropic | Orthotropic refers to a material that exhibits different properties along three mutually perpendicular axes. In the construction of a bridge deck, engineers may use orthotropic materials to enhance structural performance by tailoring material characteristics to specific load directions. (Cross-reference with Bridge Deck Design for insights into considerations specific to the design of bridge decks.)

Pendulum Mass Damper | A pendulum mass damper is a counterweight system used to mitigate vibrations in tall structures. In the design of a skyscraper, a pendulum mass damper may be incorporated to reduce swaying caused by wind or seismic events, enhancing occupant comfort and structural stability. (Cross-reference with Skyscraper Design for insights into considerations specific to tall building design.)

Quasi-Static Analysis | Quasi-static analysis involves examining the behavior of a structure under slowly applied loads. When designing a convention center with a retractable roof, engineers may use quasi-static analysis to simulate the gradual deployment of the roof structure, ensuring smooth and controlled movement. (Cross-reference with Retractable Roof Design for insights into considerations specific to retractable roof structures.)

Reinforced Concrete | Reinforced concrete is a composite material comprising concrete and embedded steel reinforcement. In the construction of a residential complex, reinforced concrete is commonly used to create durable and strong structures, combining the compressive strength of concrete with the tensile strength of steel. (Cross-reference with Residential Complex Design for insights into considerations specific to multifamily housing.)

Shear Wall | A shear wall is a vertical structural element that resists lateral forces parallel to the plane of the wall. In the design of a hotel tower, shear walls may be strategically placed to enhance the building's resistance to wind and seismic loads, ensuring stability during extreme conditions. (Cross-reference with Hotel Design for insights into considerations specific to hotel structures.)

Tensile Strength | Tensile strength is the ability of a material to withstand tension or pulling forces. In the construction of a pedestrian bridge with a cable-stayed design, engineers prioritize materials with high tensile strength to ensure the stability and safety of the bridge under dynamic loads. (Cross-reference with Pedestrian Bridge Design for insights into considerations specific to pedestrian bridge structures.)

Uplift Load | Uplift load is the force exerted on a structure in an upward direction. When designing a sports arena with a tensile fabric roof, engineers account for uplift loads to ensure that the supporting structure can withstand wind forces and maintain the integrity of the roof system. (Cross-reference with Sports Arena Design for insights into considerations specific to sports facility structures.)

Vibration Control | Vibration control involves measures to reduce unwanted oscillations in a structure. In the construction of a concert hall, engineers implement vibration control strategies to minimize the transmission of vibrations, ensuring optimal acoustics and comfort for performers and audience members. (Cross-reference with Concert Hall Design for insights into considerations specific to performance venues.)

Wind Load | Wind load is the force exerted by wind on a structure. In the design of a residential skyscraper, architects and engineers consider wind loads to optimize the building's shape and select appropriate materials, ensuring the structural integrity and comfort of occupants. (Cross-reference with Residential Skyscraper Design for insights into considerations specific to tall residential buildings.)

X-Bracing | X-bracing is a diagonal structural element used to provide lateral stability. In the construction of an office building, X-bracing may be incorporated into the structural system to resist lateral loads and enhance the overall stability of the building. (Cross-reference with Office Building Design for insights into considerations specific to office structures.)

Yield Point | Yield point is the stress at which a material undergoes permanent deformation. When designing a university laboratory building, engineers consider the yield point of structural materials to ensure that the building can withstand the loads associated with laboratory equipment and experiments. (Cross-reference with University Laboratory Design for insights into considerations specific to laboratory facilities.)

Zero Energy Building (ZEB) | A zero energy building is designed to produce as much energy as it consumes over a specified period. In the construction of a sustainable residential community, a zero energy building approach may be employed to minimize environmental impact and promote energy efficiency. (Cross-reference with Sustainable Community Design for insights into considerations specific to sustainable residential developments.)

CHAPTER 16: BUILDING ENVELOPE SYSTEMS: WALLS, ROOFS, AND WINDOWS

Air Barrier | An air barrier is a material or system that restricts the movement of air through the building envelope, preventing unwanted drafts and air leakage. In a sustainable office building, incorporating an effective air barrier helps enhance energy efficiency by reducing the need for additional heating or cooling. (Cross-reference with Sustainable Building Design for insights into energy-efficient building practices.)

Batt Insulation | Batt insulation consists of pre-cut, flexible insulation material that fits snugly between framing elements such as studs or joists. Commonly used in residential construction, batt insulation helps regulate indoor temperature by minimizing heat transfer through walls and ceilings. (Cross-reference with Residential Construction for insights into insulation applications in homes.)

Clerestory | A clerestory is a row of windows or openings near the top of a wall that allows natural light to enter a building. In the design of a modern art gallery, clerestory windows may be strategically placed to provide abundant daylighting, reducing the reliance on artificial lighting. (Cross-reference with Art Gallery Design for insights into considerations specific to gallery spaces.)

Dew Point | The dew point is the temperature at which air becomes saturated with moisture, leading to the formation of dew or condensation. Understanding the dew point is crucial in the design of a climate-controlled museum, as maintaining the interior temperature above the dew point prevents moisture-related damage to artifacts. (Cross-reference with Museum Design for insights into considerations specific to museum environments.)

Eave | An eave is the overhanging lower edge of a roof that extends beyond the exterior wall. In residential architecture, eaves play a role in protecting the building from rainwater and directing it away from the foundation. (Cross-reference with Residential Architecture for insights into architectural elements in homes.)

Fenestration | Fenestration refers to the design, arrangement, and placement of windows and other openings in a building. In a sustainable school design, careful fenestration planning maximizes natural daylight, reducing the need for artificial lighting and promoting energy efficiency. (Cross-reference with School Design for insights into considerations specific to educational facilities.)

Green Roof | A green roof is a roof covered with vegetation, providing insulation and environmental benefits. In the construction of an eco-friendly community center, a green roof may be incorporated to enhance thermal performance, reduce stormwater runoff, and contribute to biodiversity. (Cross-reference with Community Center Design for insights into considerations specific to community-oriented structures.)

Heat Transfer | Heat transfer involves the movement of thermal energy between different areas or substances. In the design of an energy-efficient library, architects consider heat transfer mechanisms to optimize insulation and HVAC systems, ensuring comfortable indoor conditions with minimal energy consumption. (Cross-reference with Library Design for insights into considerations specific to library environments.)

Insulation R-Value | The insulation R-value is a measure of insulation material's resistance to heat flow. In the construction of a sustainable residence, choosing insulation with a high R-value contributes to energy efficiency, reducing heating and cooling demands. (Cross-reference with Sustainable Residential Design for insights into sustainable practices in residential construction.)

Jamb | A jamb is the vertical portion of a door or window frame that supports the door or sash. In a healthcare facility design, specifying reinforced jambs for doors enhances security, providing durability in areas with high traffic and potential impact. (Cross-reference with Healthcare Facility Design for insights into considerations specific to healthcare environments.)

Kickout Flashing | Kickout flashing is a specialized flashing detail that prevents water from penetrating the wall at the junction of a roof and a sidewall. In the design of a sustainable hotel, incorporating kickout flashing contributes to moisture control, protecting the building envelope and enhancing longevity. (Cross-reference with Sustainable Hotel Design for insights into considerations specific to hotel structures.)

Low-E Glass | Low-emissivity (Low-E) glass is coated with a thin layer that reflects heat while allowing light to pass through. In the construction of an energy-efficient theater, the use of Low-E glass in windows contributes to climate control, minimizing heat gain and loss. (Cross-reference with Theater Design for insights into considerations specific to entertainment venues.)

Moisture Barrier | A moisture barrier is a material that prevents the passage of moisture through the building envelope. In a residential complex situated in a humid climate, a moisture barrier beneath the foundation helps protect the structure from moisture-related issues like mold and rot. (Cross-reference with Humid Climate Design for insights into considerations specific to regions with high humidity.)

Nail Fin | A nail fin is a flange used to attach windows and doors to the framing of a building. In a condominium construction project, the use of nail fins simplifies the installation process, providing a secure and weathertight connection between the windows or doors and the building structure. (Cross-reference with Condominium Construction for insights into considerations specific to multi-unit residential buildings.)

Overhang | An overhang is a projecting structure that provides shade and protection from the elements. In the design of an outdoor pavilion, incorporating overhangs contributes to user comfort by shielding occupants from direct sunlight and rainfall. (Cross-reference with Outdoor Pavilion Design for insights into considerations specific to open-air structures.)

Parapet | A parapet is a low protective wall or barrier at the edge of a roof or balcony. In the design of a mixed-use development, parapets may serve aesthetic and safety purposes, enhancing the visual appeal of the building while providing a barrier for rooftop safety. (Cross-reference with Mixed-Use Development Design for insights into considerations specific to versatile urban spaces.)

Quoins | Quoins are decorative masonry blocks or bricks used at the corners of a building. In the restoration of a historic courthouse, attention to details such as quoins preserves the architectural authenticity, contributing to the overall heritage value. (Cross-reference with Historic Courthouse Restoration for insights into considerations specific to preserving historical structures.)

Ridge Vent | A ridge vent is a ventilated feature installed along the ridge of a roof to promote air circulation. In the design of an energy-efficient warehouse, ridge vents contribute to passive ventilation, reducing the need for mechanical systems and enhancing overall energy performance. (Cross-reference with Warehouse Design for insights into considerations specific to industrial structures.)

Sill | A sill is the horizontal portion at the bottom of a window or door frame. In the design of a retail store, incorporating durable sills enhances weather resistance and longevity, ensuring the integrity of the building envelope in high-traffic commercial spaces. (Cross-reference with Retail Store Design for insights into considerations specific to commercial environments.)

Thermal Bridging | Thermal bridging occurs when a conductive material creates a pathway for heat flow, compromising insulation. In the construction of a sustainable art studio, architects address thermal bridging to minimize energy loss and maintain a comfortable working environment. (Cross-reference with Art Studio Design for insights into considerations specific to creative workspaces.)

U-Factor | The U-factor measures the rate of heat transfer through a building material or assembly. In the design of a sustainable church, selecting windows with a low U-factor contributes to energy efficiency, supporting a comfortable worship environment. (Cross-reference with Sustainable Church Design for insights into considerations specific to religious structures.)

Vapor Retarder | A vapor retarder is a material that limits the diffusion of water vapor through the building envelope. In the construction of a laboratory facility, vapor retarders are crucial in controlling moisture levels and protecting sensitive equipment. (Cross-reference with Laboratory Facility Design for insights into considerations specific to research environments.)

Weep Hole | A weep hole is a small opening in a building component that allows the drainage of water. In the design of an educational campus, incorporating weep holes in brick walls prevents water accumulation, preserving the structural integrity and appearance of the buildings. (Cross-reference with Educational Campus Design for insights into considerations specific to academic environments.)

Xeriscaping | Xeriscaping is a landscaping method that conserves water by using drought-tolerant plants. In the development of an environmentally conscious residential neighborhood, xeriscaping contributes to sustainable water management and creates a visually appealing landscape. (Cross-reference with Sustainable Residential Neighborhood Design for insights into considerations specific to eco-friendly communities.)

Yard Drainage | Yard drainage involves the design and installation of systems to manage water runoff in outdoor spaces. In the planning of a recreational park, effective yard drainage ensures a safe and well-maintained environment, preventing waterlogging and promoting usability. (Cross-reference with Recreational Park Design for insights into considerations specific to outdoor recreational areas.)

Zinc Roofing | Zinc roofing involves the use of zinc as a material for roofing applications. In the construction of a modern airport terminal, zinc roofing may be chosen for its durability and resistance to corrosion, providing long-lasting protection against the elements. (Cross-reference with Airport Terminal Design for insights into considerations specific to transportation hubs.)

CHAPTER 17: MECHANICAL, ELECTRICAL, PLUMBING, AND FIRE PROTECTION SYSTEMS

Air Handling Unit (AHU): An AHU is a crucial component of HVAC systems, responsible for circulating and regulating air. In a sustainable office building, the AHU's energy-efficient design ensures optimal air quality and thermal comfort, contributing to overall environmental performance.

Backup Power System: A backup power system, like a generator or UPS, provides electricity during outages. In a hospital setting, the backup power system ensures the continuous operation of critical medical equipment, emphasizing its role in maintaining patient care.

Circuit Breaker: Circuit breakers protect electrical circuits from overloads or faults. In an industrial facility, the proper selection and maintenance of circuit breakers contribute to electrical safety and prevent disruptions in manufacturing processes.

Ductwork: Ductwork is a network of channels that distribute air throughout a building. In a high-rise residential complex, well-designed ductwork ensures efficient heating and cooling, enhancing indoor comfort for residents.

Emergency Lighting: Emergency lighting systems provide illumination during power outages. In a hotel, emergency lighting is strategically placed to guide guests safely to exits during evacuations, emphasizing its role in life safety.

Fire Sprinkler System: Fire sprinkler systems are crucial for suppressing fires in buildings. In a museum housing valuable artifacts, a well-designed sprinkler system, combined with fire-resistant materials, plays a key role in preserving cultural treasures.

Generator: Generators provide backup power, especially in areas prone to outages. In a data center, generators ensure uninterrupted operation, preventing data loss and maintaining critical information systems.

HVAC (Heating, Ventilation, and Air Conditioning): HVAC systems control indoor temperature and air quality. In a modern office building, a well-balanced HVAC system promotes a healthy and productive work environment, contributing to employee well-being.

Insulation (MEP Systems): Proper insulation is essential for energy efficiency. In a residential development, effective insulation in MEP systems reduces heating and cooling demands, aligning with sustainable construction practices.

Junction Box: Junction boxes enclose electrical connections. In a shopping mall, junction boxes facilitate the safe distribution of power for lighting and other electrical systems, ensuring a well-lit and secure environment.

Kilowatt-hour (kWh): Kilowatt-hour is a unit of energy

consumption. In a smart home, monitoring kWh usage helps homeowners optimize energy use, contributing to energy efficiency and cost savings.

Lighting Control System: Lighting control systems regulate the intensity and timing of illumination. In a convention center, a sophisticated lighting control system adapts to different events, enhancing visual experiences and conserving energy.

Mains (Electrical): Mains refer to the primary power supply. In an educational institution, maintaining reliable electrical mains is essential for uninterrupted teaching and learning activities, highlighting its importance in institutional continuity.

Natural Gas Piping: Natural gas piping delivers gas for various applications. In a restaurant kitchen, natural gas piping fuels stoves and ovens, emphasizing its role in supporting culinary operations.

Occupancy Sensor: Occupancy sensors detect human presence and adjust lighting accordingly. In a commercial office space, occupancy sensors contribute to energy efficiency by automatically turning off lights in unoccupied areas.

Plumbing Fixtures: Plumbing fixtures include sinks, faucets, and toilets. In a luxury hotel, high-quality plumbing fixtures enhance guest experiences, reflecting the establishment's commitment to comfort and aesthetics.

Quick-Response Sprinkler: Quick-response sprinklers activate rapidly to control fires. In a warehouse storing flammable goods, the installation of quick-response sprinklers is critical for swift and effective fire suppression.

Refrigerant: Refrigerants are substances used in cooling systems. In a supermarket, environmentally friendly refrigerants contribute to sustainable refrigeration practices, aligning with global efforts to reduce greenhouse gas emissions.

Solar Photovoltaic System: Solar PV systems harness sunlight to generate electricity. In a residential community, the integration of solar photovoltaic systems reduces dependency on the grid, promoting renewable energy adoption and sustainability.

Transformer: Transformers adjust voltage levels in electrical systems. In an industrial facility, transformers ensure the efficient distribution of electricity, supporting manufacturing processes and equipment.

UPS (Uninterruptible Power Supply): UPS systems provide temporary power during outages. In a financial institution, UPS systems safeguard critical data and operations, preventing financial losses due to power disruptions.

Valve (Plumbing): Plumbing valves control the flow of water. In a high-rise apartment building, well-designed plumbing valves ensure water efficiency and emergency shut-off capabilities, contributing to overall water management.

Water Heater: Water heaters provide hot water for various applications. In a fitness center, efficient water heaters support amenities like showers and spa facilities, enhancing the overall customer experience.

Xenon Lighting: Xenon lighting offers bright and focused illumination. In a theater, xenon lighting enhances stage visibility and color rendition, contributing to captivating performances.

Yard Hydrant: Yard hydrants provide water access in outdoor areas. In an industrial complex, strategically placed yard hydrants support firefighting efforts, ensuring rapid response and protection against potential hazards.

Zone Control (HVAC): HVAC zone control allows tailored temperature management in different areas. In a multi-use commercial building, zone control optimizes energy use by adjusting heating and cooling based on specific occupancy patterns.

CHAPTER 18:
BUILDING FINISHING
AND INTERIORS

Acoustic Panel: Acoustic panels are specialized materials designed to absorb sound and reduce noise levels within a space. In a home theater, acoustic panels may be strategically placed to enhance audio quality by minimizing echoes and reverberations. These panels come in various designs and materials, allowing interior designers to balance aesthetics with acoustic performance. The selection and placement of acoustic panels contribute to creating an immersive and acoustically optimized environment.

Baseboard: A baseboard, also known as a skirting board, is a molding installed at the base of an interior wall. In a contemporary living room, baseboards can add a finishing touch to the transition between walls and floors. Beyond aesthetics, baseboards serve a functional purpose by protecting the base of the wall from damage. Interior designers coordinate the design and color of baseboards to complement the overall style of a room.

Crown Molding: Crown molding is a decorative trim installed along the top of interior walls, where the walls meet the ceiling. In a classic dining room, crown molding can add an elegant and architectural detail to the space. The design and intricacy of crown molding contribute to the overall aesthetic of a room. Interior designers may use crown molding to create visual interest and define the character of different areas within a home.

Decorative Hardware: Decorative hardware refers to stylish and visually appealing elements such as handles, knobs, and pulls used on cabinets, doors, and furniture. In a luxury kitchen, decorative hardware can enhance the overall design theme and add a touch of sophistication. Interior designers pay attention to the material, finish, and design of decorative hardware to ensure cohesiveness with the overall design concept.

Epoxy Flooring: Epoxy flooring involves the application of a durable and seamless resin coating over existing flooring surfaces. In a modern commercial space, epoxy flooring may be chosen for its aesthetic appeal, durability, and ease of maintenance. Interior designers consider the color and texture options available with epoxy flooring to create visually striking and resilient floor surfaces.

Faux Finish: Faux finish is a decorative painting technique that replicates the appearance of materials such as wood, marble, or stone. In a contemporary bedroom, faux finishing may be applied to create a textured accent wall with the look of aged stone. Interior designers use faux finish techniques to add depth and visual interest to surfaces, achieving a desired aesthetic without the use of expensive materials.

Grout: Grout is a cementitious material used to fill the gaps between tiles and provide stability to the tiled surface. In a spa-inspired bathroom, the choice of grout color can significantly impact the overall look of the tiled area. Interior designers consider grout as an important design element, selecting colors that complement or contrast with the tiles to achieve the desired visual effect.

Hardwood Flooring: Hardwood flooring involves the use of natural wood planks to create durable and visually appealing floor surfaces. In a classic living room, hardwood flooring can add warmth and timeless elegance. Interior designers choose hardwood species, finishes, and plank sizes based on the design theme and desired atmosphere of the space.

Interior Design: Interior design encompasses the art and science of enhancing the interior of a space to create a functional and aesthetically pleasing environment. In a contemporary office, interior design principles are applied to optimize the layout, furniture selection, and overall ambiance. Interior designers consider factors such as color, lighting, and spatial arrangement to achieve a cohesive and inviting interior.

Jamb: A jamb refers to the vertical sides of a door or window frame. In a modern entryway, the design of door jambs can contribute to the overall architectural style. Interior designers may select door jambs with specific profiles and finishes to align with the design theme of a space.

Kickplate: A kickplate is a protective panel installed at the bottom of a door to prevent damage from foot traffic. In a high-traffic commercial space, kickplates may be used to safeguard the base of doors from scuff marks and scratches. Interior designers choose kickplate materials and finishes that balance durability with visual appeal, enhancing the longevity of doors in busy areas.

Laminate Flooring: Laminate flooring consists of synthetic layers that mimic the appearance of wood or other materials. In a contemporary kitchen, laminate flooring can provide a cost-effective and versatile alternative to traditional hardwood. Interior designers explore a variety of laminate designs and textures to achieve the desired aesthetic while considering practical factors such as maintenance and durability.

Molding: Molding refers to decorative trim or edging used to enhance the visual appeal of surfaces, such as walls, ceilings, or furniture. In a luxury dining room, intricate molding details can be applied to create a sense of grandeur. Interior designers skillfully incorporate molding elements to define architectural features and elevate the overall design sophistication of a space.

Niche: A niche refers to a shallow recess, often in a wall, designed to display or highlight objects. In a contemporary residence, niches can be strategically incorporated to showcase art pieces or sculptures, adding a personalized touch to interior spaces. Beyond aesthetics, niches can serve practical purposes, such as providing storage or serving as a decorative focal point. Interior designers leverage niches to enhance spatial dynamics and create unique points of interest within a room.

Oriel Window: An oriel window is a type of projecting window that extends from the main structure of a building. In a historic library, oriel windows may be incorporated to provide reading nooks with ample natural light. The design of oriel windows can be customized to align with architectural styles, contributing to the overall character of a building. Interior designers may strategically position furnishings around oriel windows to optimize views and create inviting spaces.

Partition: A partition is a structure or barrier that divides a space into distinct areas. In a contemporary office layout, glass partitions may be used to create separate workspaces while maintaining a sense of openness. The choice of materials and design for partitions influences both visual aesthetics and functional aspects of a space. Interior designers consider partitions as essential elements for optimizing spatial flow and creating purposeful zones within larger areas.

Quoins: Quoins are decorative masonry blocks or bricks that emphasize the corners of a building. In a luxury hotel, quoins may be incorporated to accentuate the architectural features and provide a sense of grandeur. The design and material selection for quoins contribute to the overall aesthetic harmony of a structure. Interior designers working on upscale projects may coordinate the color and texture of quoins with other design elements to create a cohesive visual narrative.

Riser: In the context of stairs, a riser is the vertical component that connects consecutive steps. In a modern residential setting, custom-designed risers with intricate patterns can transform a staircase into a striking focal point. The selection of materials and finishes for risers contributes to the overall design theme, whether it's a minimalist approach or a more ornate style. Interior designers pay attention to riser details to ensure cohesiveness with the broader design scheme of a space.

Skirting Board: A skirting board, also known as a baseboard, is a molding that runs along the base of an interior wall. In a contemporary retail store, skirting boards may be selected to complement the overall design aesthetic and color palette. While offering a finishing touch to walls, skirting boards serve both decorative and protective purposes. Interior designers consider the visual impact of skirting boards on the overall design scheme, ensuring harmony with other design elements.

CHAPTER 19:
BUILDING CODES
AND REGULATIONS

Accessibility Standards: Accessibility standards are crucial guidelines ensuring that buildings and facilities are designed to accommodate individuals with disabilities, promoting inclusivity. For instance, a ramp installation (cross-reference with Egress) in compliance with accessibility standards provides wheelchair users with a safe means of egress during emergencies, aligning with regulatory requirements.

Building Code: A comprehensive set of regulations governing construction, materials, safety, and occupancy, building codes are enforced to guarantee the well-being of occupants. Compliance with building codes ensures the construction of structurally sound buildings; for example, adherence to seismic provisions in earthquake-prone areas (cross-reference with Structural Systems and Design for Buildings).

Certificate of Occupancy: This official document, issued by local authorities, attests that a building complies with building codes and is fit for occupancy. A residential building meeting green building codes may receive a certificate of occupancy, showcasing its sustainable features (cross-reference with Green Building Codes).

Deed Restrictions: Restrictions outlined in property deeds that dictate land use. An architectural design adhering to specific aesthetics might be influenced by deed restrictions, ensuring harmony with the neighborhood (cross-reference with Aesthetics).

Egress: Egress refers to the safe exit from a building during emergencies, encompassing features like exit doors, corridors, and stairways. An egress plan may incorporate fire doors (cross-reference with Fire Code) designed to resist fire and contribute to a safe evacuation.

Fire Code: Fire codes, a subset of building codes, focus on preventing and mitigating fire risks. For instance, the installation of fire-resistant materials (cross-reference with Construction Materials) is a key element in fire code compliance, enhancing building safety.

Green Building Codes: These codes promote environmentally sustainable construction practices, emphasizing energy efficiency and eco-friendly materials. Incorporating solar panels (cross-reference with Sustainable Construction Practices) aligns with green building codes, contributing to energy efficiency.

Historic Preservation Guidelines: Standards for preserving and protecting historic structures, these guidelines influence renovations and alterations. When renovating a historic building, adherence to preservation guidelines may involve using period-appropriate materials (cross-reference with Construction Materials).

Inspection: The examination of buildings to ensure compliance with codes and regulations. For example, an inspection may verify that electrical systems meet the National Electrical Code (NEC) standards (cross-reference with NEC).

Jurisdictional Approvals: Necessary permissions from local authorities before construction commences. Obtaining permits for specialized work, such as electrical installations (cross-reference with Electrical Systems), is a crucial aspect of jurisdictional approvals.

Zoning Regulations: Local rules defining land use within a jurisdiction, zoning regulations impact property development. Compliance with zoning regulations may influence the layout of a site plan (cross-reference with Site Planning).

Land Use Controls: Measures regulating land use for sustainable development. Land use controls, including zoning regulations, guide decisions related to construction materials (cross-reference with Construction Materials) to ensure environmental responsibility.

Materials Testing: Analysis of construction materials to ensure compliance with standards. Testing materials for durability (cross-reference with Durability) ensures that structures meet performance expectations over time.

National Electrical Code (NEC): A standard for safe electrical installations, the NEC is crucial for electrical systems in buildings. Compliance with the NEC is vital for ensuring the safety of electrical wiring and equipment.

Occupancy Classification: Categorizing buildings based on primary use, occupancy classification determines applicable code requirements. For example, a building designated for educational use may require specific fire safety features (cross-reference with Fire Code).

Performance Standards: Criteria assessing the performance of buildings and systems. High-performance windows (cross-reference with Building Envelope Systems) may contribute to energy efficiency, aligning with performance standards.

Qualified Inspector: An individual certified to conduct inspections, a qualified inspector ensures adherence to codes and standards. A qualified inspector may assess the structural integrity of a building (cross-reference with Structural Systems and Design for Buildings).

Regulatory Compliance: Adherence to laws, codes, and regulations governing construction. Regulatory compliance may involve incorporating stormwater management practices (cross-reference with Site Planning) to meet environmental regulations.

Special Permits: Permits allowing specific uses or deviations from standard regulations. Obtaining a special permit for unique architectural features (cross-reference with Aesthetics) may involve demonstrating their compatibility with the overall design.

Trade Permits: Permits for specific trades within a construction project. For example, obtaining a plumbing permit (cross-reference with Plumbing Systems) is essential for ensuring compliance with plumbing codes.

Use Variance: Permission for non-standard use granted under specific circumstances. Obtaining a use variance may involve demonstrating that a proposed use aligns with community aesthetics (cross-reference with Aesthetics).

Variance: An exception granted from standard regulations. Variances may be sought for setbacks (cross-reference with Yard Setbacks) to accommodate unique site conditions.

Warranty Period: The duration during which a builder is responsible for addressing defects. During the warranty period, a builder may address issues related to construction materials (cross-reference with Construction Materials) to ensure long-term durability.

Xeriscaping (Exterior): Landscaping practices for water efficiency, xeriscaping aligns with environmental regulations. Utilizing drought-resistant plants (cross-reference with Landscape Design) complies with xeriscaping principles.

Yard Setbacks: Required distances between buildings and property lines, yard setbacks impact site planning. Compliance with yard setbacks may influence the placement of outdoor amenities (cross-reference with Landscape Design).

Zoning Board of Appeals: A board addressing appeals and variances related to zoning regulations. A zoning board of appeals may grant variances for architectural features (cross-reference with Aesthetics) that deviate from standard regulations.

CHAPTER 20: BUILDING SUSTAINABILITY AND ENERGY EFFICIENCY

Air Infiltration: The unintended flow of air into and out of a building, air infiltration affects energy efficiency. Addressing air infiltration may involve using weatherstripping (cross-reference with Weatherstripping) to seal gaps and improve insulation.

Biomass: Organic materials used as fuel, biomass contributes to sustainable energy sources. Incorporating biomass heating systems (cross-reference with Heating Systems) aligns with sustainable practices, utilizing renewable resources.

Carbon Footprint: The total amount of greenhouse gases, primarily carbon dioxide, emitted directly or indirectly by an individual, organization, event, or product. Reducing the carbon footprint may involve implementing energy-efficient systems (cross-reference with Energy Efficiency).

Daylighting: The practice of utilizing natural daylight to illuminate the interior of a building, daylighting enhances energy efficiency. Designing spaces with ample windows and skylights (cross-reference with Skylights) promotes effective daylighting.

Energy Efficiency: The use of technology and practices to reduce energy consumption without sacrificing performance. Installing energy-efficient lighting systems (cross-reference with Lighting Systems) is a key strategy for enhancing overall energy efficiency.

Fenestration: The arrangement, proportioning, and design of windows and doors in a building. Selecting energy-efficient fenestration systems, such as double-glazed windows (cross-reference with Windows), contributes to sustainable design.

Geothermal Energy: Renewable energy derived from the heat of the Earth's interior, geothermal energy offers a sustainable heating and cooling solution. Implementing a geothermal heat pump system (cross-reference with Heating Systems) harnesses geothermal energy for efficient temperature control.

Heat Recovery Ventilation (HRV): A mechanical ventilation system that recovers heat from outgoing air and uses it to preheat incoming air. Incorporating HRV systems (cross-reference with Ventilation Systems) enhances energy efficiency by reducing the need for additional heating.

Insulation: Materials used to reduce heat transfer, insulation is critical for energy-efficient buildings. Choosing eco-friendly insulation materials (cross-reference with Construction Materials) contributes to both sustainability and energy efficiency.

LEED Certification: Leadership in Energy and Environmental Design (LEED) certification is a globally recognized rating system for sustainable building design, construction, and operation. Achieving LEED certification involves incorporating sustainable features, such as green roofs (cross-reference with Roofing Systems).

Microgrid: A localized energy system capable of operating independently or in conjunction with the main power grid.

Implementing a microgrid with renewable energy sources (cross-reference with Renewable Energy) enhances resilience and sustainability.

Net Zero Energy Building: A building designed to produce as much energy as it consumes over the course of a year. Achieving net-zero status may involve incorporating solar panels (cross-reference with Solar Panels) and energy-efficient systems.

Occupancy Sensors: Devices that detect the presence or absence of people and adjust lighting and HVAC systems accordingly. Installing occupancy sensors in spaces (cross-reference with Lighting Systems) helps optimize energy usage by turning off systems when not needed.

Passive Design: Design strategies that leverage natural elements, such as sunlight and wind, to enhance energy efficiency. Incorporating passive design elements, such as natural ventilation (cross-reference with Ventilation Systems), reduces reliance on active systems.

Quality of Indoor Environment (QIE): An assessment of the indoor environment's impact on occupant health and well-being. Enhancing QIE may involve implementing green building practices (cross-reference with Green Building Codes) that prioritize occupant comfort.

Renewable Energy: Energy derived from sources that are naturally replenished, such as sunlight, wind, and hydropower. Integrating renewable energy systems, such as wind turbines (cross-reference with Wind Turbines), supports sustainable and low-carbon energy production.

Solar Panels: Photovoltaic panels that convert sunlight into electricity. Installing solar panels on rooftops (cross-reference with Roofing Systems) provides a sustainable and renewable energy source for buildings.

Triple Bottom Line (TBL): An accounting framework that

considers social, environmental, and economic factors in decision-making. Designing with the TBL in mind may involve incorporating sustainable construction materials (cross-reference with Construction Materials) that align with environmental goals.

Utility Rebates: Financial incentives offered by utility companies to encourage energy-efficient practices. Implementing energy-efficient systems (cross-reference with Energy Efficiency) may qualify a building for utility rebates, promoting cost savings.

Ventilation Systems: Mechanical systems that provide fresh air to indoor spaces. Integrating energy-efficient ventilation systems, such as demand-controlled ventilation (cross-reference with Demand-Controlled Ventilation), improves indoor air quality while minimizing energy consumption.

Water Conservation: Practices and technologies aimed at reducing water consumption. Implementing water-saving fixtures (cross-reference with Plumbing Systems) contributes to sustainable water usage in buildings.

Xeriscaping (Exterior): Landscaping practices for water efficiency, xeriscaping aligns with sustainable water use. Incorporating xeriscaping in exterior spaces (cross-reference with Landscape Design) minimizes water consumption.

Yield: The energy output produced by renewable energy systems. Calculating the yield of a solar energy system (cross-reference with Solar Panels) provides insights into its efficiency and contribution to energy needs.

Zero Net Energy (ZNE): Similar to a net-zero energy building, a Zero Net Energy (ZNE) project balances the amount of energy consumed with an equivalent amount of renewable energy production. Achieving ZNE status involves comprehensive energy management and renewable energy integration.

PART IV:
CONSTRUCTION PROJECT DELIVERY AND PROCUREMENT

CHAPTER 21: PROJECT DELIVERY METHODS: TRADITIONAL, DESIGN-BUILD, AND OTHERS

Bid-Build: A traditional project delivery method where the owner contracts with separate entities for design and construction. The project goes through a bidding process, and the selected contractor builds based on the completed design.

Bridging: A design-build project delivery method that involves hiring a design professional to develop initial project documents before selecting a design-build team. This helps set a baseline for the design-build team.

CM at Risk (Construction Management at Risk): A delivery method where the construction manager is engaged during the design phase to provide input on cost and constructability. The construction manager then transitions to the role of a general contractor during construction.

CMa (Construction Manager as Advisor): In this project delivery method, the construction manager acts as an advisor to the owner during the design phase. The owner contracts directly with trade contractors during construction.

Design-Bid-Build (DBB): The traditional project delivery method where the owner contracts separately with a designer and a construction contractor. Design is completed before the project goes out to bid.

Integrated Project Delivery (IPD): A collaborative project delivery method where all stakeholders, including the owner, designer, and contractor, work together from the project's early stages. Risks and rewards are shared among the team.

Job Order Contracting (JOC): A project delivery method where contractors are pre-selected through competitive bidding. The selected contractor then provides construction services on an as-needed basis through a set of unit prices.

Multiple-Prime Contracts: A project delivery method where the owner contracts directly with multiple prime contractors for different aspects of the project. Each prime contractor is responsible for a specific scope of work.

PPP (Public-Private Partnership): A project delivery method where a private entity partners with a public entity to finance, design, build, operate, and maintain a public infrastructure project.

Progressive Design-Build: A design-build project delivery method where the owner selects a design-build team based on qualifications, and the design evolves collaboratively as the project progresses.

Public-Private Partnership (PPP): A project delivery method where a private entity partners with a public entity to finance, design, build, operate, and maintain a public infrastructure project.

Traditional Project Delivery: A method where the owner contracts separately with a designer and a construction contractor. Design is completed before the project goes out to bid.

Turnkey: A design-build project delivery method where the design-build team is responsible for the entire project, from design to construction and sometimes even financing.

CHAPTER 22: CONSTRUCTION PROCUREMENT AND BIDDING STRATEGIES

Auction: A procurement strategy where bidders compete openly by submitting successively higher bids until the highest bidder is accepted.

Best Value Procurement: A procurement strategy that considers not only cost but also other factors such as quality, performance, and past experience when selecting a contractor.

Competitive Sealed Bidding: A procurement method where bids are submitted in sealed envelopes and opened publicly. The contract is awarded to the lowest responsive and responsible bidder.

Construction Manager at Risk (CM at Risk): A procurement strategy where the construction manager is engaged during the design phase to provide input on cost and constructability. The construction manager then transitions to the role of a general contractor during construction.

Design-Build: A procurement method where the design and construction services are provided by a single entity, the design-build team, which can streamline the process and promote collaboration.

Electronic Reverse Auction: An online auction where suppliers bid against each other to provide goods or services at the lowest price.

Fixed-Price Contract: A type of contract where the contractor agrees to perform the work for a predetermined, fixed price, providing cost certainty to the owner.

Guaranteed Maximum Price (GMP) Contract: A contract where the construction manager at risk guarantees that the project will not exceed a specified maximum price, providing some cost certainty to the owner.

Invitation for Bid (IFB): A formal invitation to contractors to submit sealed bids for a construction project following a competitive sealed bidding process.

Joint Ventures: An arrangement where two or more contractors join forces to submit a bid for a construction project, combining their resources and expertise.

Letter of Intent (LOI): A document expressing an intent to enter into a contract, often used in the preliminary stages of procurement to secure a contractor's commitment.

Negotiated Procurement: A procurement method where the owner negotiates directly with a contractor to determine the contract terms, including price and scope of work.

Prequalification: The process of evaluating and selecting contractors based on their qualifications, experience, and financial capabilities before inviting them to bid on a project.

Request for Proposals (RFP): A document inviting contractors to submit proposals for a project, often used in design-build or other collaborative procurement methods.

Single-Source Procurement: A procurement strategy where the owner selects a single contractor without competitive bidding, often justified by unique expertise or a specialized scope of work.

Time and Materials (T&M) Contract: A type of contract where the contractor is paid based on the time spent and materials used, suitable for projects with uncertain or evolving requirements.

Unit Price Contract: A contract where the contractor is paid based on the quantity of work performed at predetermined unit prices, providing flexibility for varying project conditions.

Value Engineering: A systematic process of reviewing a project to identify opportunities to achieve the same or better results at a lower cost.

Weighted Criteria: A method of evaluating bids where different factors are assigned different weights, allowing the owner to consider both cost and other qualitative criteria.

Xtreme Source Selection: An innovative procurement method (hypothetical term for illustration purposes) that combines elements of various procurement strategies to optimize project outcomes.

Yield-Premium Bidding: A bidding strategy where contractors bid not only on the base project but also on alternate items, allowing the owner to choose additional work based on available funds.

Zero-Bid: A situation where no acceptable bids are received for a construction project, leading to the need for reevaluation or re-bidding.

CHAPTER 23: CONTRACT ADMINISTRATION AND DISPUTE RESOLUTION

Acceptance: The formal approval or acknowledgement of the completed work or deliverables by the owner or the designated authority.

Breach of Contract: A violation of the terms and conditions specified in a contract, which may result in legal consequences or remedies.

Change Order: A formal document that modifies the scope, schedule, or cost of a construction project, usually initiated by the owner or contractor to address changes in the project's requirements.

Dispute Resolution: The process of resolving disagreements or conflicts that may arise during the course of a construction project. This can include mediation, arbitration, or litigation.

Earnest Money: A deposit made by the bidder as a sign of good faith when submitting a bid. It is typically a small percentage of the bid amount.

Force Majeure: Unforeseeable circumstances that prevent a party from fulfilling their contractual obligations, such as natural disasters or other unavoidable events.

Guarantee/Warranty Period: The duration during which the contractor is responsible for correcting any defects or issues that arise after the completion of the construction project.

Hold Harmless Clause: A contractual provision in which one party agrees to indemnify and protect the other party from any losses or claims.

Indemnification: The act of compensating or reimbursing one party for potential losses or damages incurred, often specified in a contract.

Joint Check Agreement: An agreement where a check is made jointly payable to two parties, typically the contractor and a subcontractor or supplier, ensuring that both parties are paid.

Key Performance Indicators (KPIs): Metrics used to measure the success or performance of a construction project. These indicators can include cost performance, schedule adherence, and quality standards.

Liquidated Damages: A predetermined amount of money specified in a contract that a party must pay if they fail to meet certain contractual obligations, often related to project delays.

Milestone: A significant point or event in a construction project that indicates the completion of a specific phase or task.

No-Damage-for-Delay Clause: A contractual provision that limits the contractor's ability to claim damages for project delays caused by the owner or other factors beyond the contractor's control.

Overhead: Indirect costs incurred by a contractor that are not directly tied to a specific construction project but are necessary for the overall operation of the business.

Punch List: A list of minor tasks, defects, or incomplete items that need to be addressed or corrected before a construction project is considered complete.

Quality Assurance: The processes and procedures implemented to ensure that the construction project meets specified quality standards.

Retention: A portion of the contract price withheld by the owner until the completion of the construction project to ensure that the contractor fulfills all contractual obligations.

Substantial Completion: The point in a construction project when the work is sufficiently complete, allowing the owner to occupy or use the project for its intended purpose.

Termination for Convenience: A contractual provision that allows either party to terminate the contract without default when it is in their best interest to do so.

Utility Easement: A right granted by a property owner to allow another party, such as a utility company, to use a specific portion of the property for certain purposes.

Value Engineering Change Proposal (VECP): A proposal submitted by a contractor suggesting changes to the project that may result in cost savings while maintaining or improving performance.

Workmanship: The quality of the craftsmanship and skill applied to the construction project, ensuring that the work is performed to the required standards.

CHAPTER 24: RISK MANAGEMENT IN CONSTRUCTION PROJECTS

Allowance: In construction, an allowance refers to a budgeted amount set aside to cover potential costs or activities that cannot be precisely determined during the initial project estimation. This serves as a contingency fund to address unforeseen circumstances, such as changes in material prices or unexpected site conditions. For instance, if a construction project involves excavating the foundation, an allowance might be allocated for potential discoveries of unforeseen obstacles beneath the soil, such as buried debris.

Bidding Risks: Bidding risks encompass uncertainties associated with the competitive bidding process in construction projects. This may include fluctuations in market conditions, the accuracy of bid estimates, and the competitiveness of other bids. An example of bidding risks is when contractors submit bids based on assumptions about material costs, and if these costs rise unexpectedly, it may lead to financial challenges. Cross-referencing with "Financial Risks" is pertinent, as bidding risks can directly impact the financial aspects of a construction project.

Contingency Planning: Contingency planning involves developing strategies and actions to respond to potential risks or unforeseen events during the construction project. For example, if there's a contingency plan in place for adverse weather conditions during a critical construction phase, it may include scheduling buffers or alternative work plans to mitigate the impact on the project timeline. Contingency planning is closely related to "Risk Management Plan," as it is an integral component of proactive risk management.

Design Risks: Design risks are associated with the design phase of a construction project and involve potential errors, omissions, or changes in design requirements. An example is when modifications to the original design are needed due to unforeseen site conditions. Cross-referencing with "Change Order" is relevant, as design risks may lead to changes in the project scope, triggering the need for formal change orders.

Environmental Risks: Environmental risks pertain to the potential impact of environmental factors on a construction project. This includes considerations such as weather conditions, geological issues, or regulatory changes. For instance, a construction project located in a flood-prone area faces environmental risks related to potential flooding. Cross-referencing with "Geotechnical Risks" is important, as both categories involve factors related to the project site and surroundings.

Financial Risks: Financial risks are related to the project's financial aspects, encompassing challenges such as cost overruns, funding shortages, or fluctuations in interest rates. An example is when unexpected costs arise during construction, affecting the overall project budget. Cross-referencing with "Bidding Risks" is significant, as financial risks are often intertwined with uncertainties arising from the bidding process.

Geotechnical Risks: Geotechnical risks are associated with the ground conditions at the construction site, including soil stability, subsurface conditions, and geological factors. An example is encountering unexpectedly soft soil that may require additional foundation support. Cross-referencing with "Environmental Risks" is relevant, as both categories involve considerations related to the project site.

Hazard Identification: Hazard identification is the process of identifying potential hazards or risks that could adversely affect the construction project. This includes recognizing elements such as unsafe work conditions or materials. An example is identifying potential fall hazards on a construction site and implementing safety measures. Cross-referencing with "Job Safety Analysis (JSA)" is logical, as both concepts relate to ensuring a safe working environment through risk assessment.

Insurance Coverage: Insurance coverage involves policies and options to protect against specific risks, including liability, property damage, and workers' compensation. An example is a construction liability insurance policy that provides coverage for accidents or injuries on the construction site. Cross-referencing with "Risk Management Plan" is essential, as insurance coverage is a key component of a comprehensive risk management strategy.

Job Safety Analysis (JSA): Job Safety Analysis is a systematic procedure to identify, assess, and control potential hazards associated with specific job tasks on the construction site. An example is conducting a JSA before working with heavy machinery to identify and mitigate potential safety risks. Cross-referencing with "Hazard Identification" is logical, as both concepts contribute to creating a safe work environment by analyzing and addressing potential risks.

Key Risk Indicators (KRIs): Key Risk Indicators are metrics or indicators used to monitor and measure potential risks throughout the construction project. An example is tracking indicators such as cost variances or schedule deviations to proactively identify emerging risks. Cross-referencing with "Risk Management Plan" is pertinent, as KRIs are integral to an effective risk management strategy.

Life-Cycle Risks: Life-cycle risks are risks that may arise at different stages of the construction project life cycle, including planning, design, construction, and operation. An example is recognizing that certain materials chosen during the design phase may pose maintenance challenges in the project's later stages. Cross-referencing with "Project Life Cycle" is important, as life-cycle risks span the entire duration of a construction project.

Material Procurement Risks: Material procurement risks are associated with the acquisition of construction materials, encompassing challenges such as delays, shortages, or quality issues. An example is a construction project facing delays due to the late delivery of critical materials. Cross-referencing with "Supply Chain Risks" is relevant, as material procurement risks are often tied to uncertainties in the supply chain.

Negotiated Risk: Negotiated risk refers to risks that are explicitly addressed and allocated through negotiation and agreement among project stakeholders. An example is a project team negotiating the allocation of risks between the owner and contractor in a construction contract. Cross-referencing with "Contractual Risk Allocation" is logical, as negotiated risks often involve contractual agreements defining responsibilities and liabilities.

Owner's Risk Tolerance: Owner's risk tolerance represents the level of risk that the project owner is willing to accept, influencing risk management decisions and strategies. An example is an owner with a low risk tolerance prioritizing conservative project planning and contingency measures. Cross-referencing with "Stakeholder Risk Preferences" is significant, as owner's risk tolerance aligns with broader stakeholder considerations.

Project Risk Management Plan: A Project Risk Management Plan is a comprehensive plan that outlines how risks will be identified, assessed, and managed throughout the construction project. An example is a detailed document specifying risk assessment methodologies, mitigation strategies, and contingency plans. Cross-referencing with "Contingency Planning" is essential, as the risk management plan incorporates contingency measures to address unforeseen events.

Quality Risks: Quality risks are associated with the potential for defects, non-compliance with specifications, or other issues affecting the quality of construction work. An example is the risk of using substandard materials that may compromise the structural integrity of a building. Cross-referencing with "Quality Management" is pertinent, as both concepts relate to maintaining and ensuring the quality of construction projects.

Regulatory Risks: Regulatory risks arise from changes in regulations, permits, or compliance requirements that may impact the construction project. An example is a construction project facing delays due to changes in zoning regulations. Cross-referencing with "Environmental Risks" is logical, as both categories involve external factors that may influence project requirements.

Schedule Risks: Schedule risks are related to potential delays or disruptions in the project schedule, including unforeseen events or changes in scope. An example is a construction project experiencing delays due to inclement weather conditions. Cross-referencing with "Critical Path" is relevant, as schedule risks often involve critical activities that can impact the overall project timeline.

Technology Risks: Technology risks are associated with the use of technology, software, or innovative construction methods that may introduce uncertainties. An example is adopting new building information modeling (BIM) software without proper training, leading to errors in project documentation. Cross-referencing with "Innovation Risks" is logical, as technology risks are a subset of broader risks associated with innovative practices.

Uncertainty Analysis: Uncertainty analysis involves the assessment of uncertainties and unknowns in project parameters to better understand and manage potential risks. An example is using probabilistic models to evaluate the range of possible project outcomes based on different scenarios. Cross-referencing with "Risk Assessment" is significant, as both concepts involve evaluating and quantifying uncertainties.

Value at Risk (VaR): Value at Risk is a quantitative measure representing the potential financial loss associated with specific risks in a construction project. An example is calculating the VaR to estimate the financial impact of a worst-case scenario, helping project stakeholders make informed decisions. Cross-referencing with "Financial Risks" is pertinent, as VaR is a financial metric closely tied to potential monetary losses.

Weather Risks: Weather risks are associated with adverse weather conditions, including rain, snow, hurricanes, or extreme temperatures. An example is a construction project in a region prone to hurricanes facing weather-related delays. Cross-referencing with "Force Majeure" is logical, as adverse weather conditions are often considered force majeure events in construction contracts.

X-factor Risks: X-factor risks are unforeseen or unpredictable risks that may not fit into conventional risk categories but have the potential to impact the project. An example is a sudden economic downturn affecting project financing unexpectedly.

Cross-referencing with "Black Swan Events" is relevant, as X-factor risks often share characteristics with rare and unpredictable events.

Yield Analysis: Yield analysis involves the evaluation of potential returns or benefits relative to the level of risk in a construction project. An example is assessing the expected yield of a real estate development project compared to the associated risks. Cross-referencing with "Return on Investment (ROI)" is logical, as both concepts involve evaluating project outcomes against associated risks.

Zero Tolerance: Zero tolerance in construction refers to the approach of having no tolerance for specific risks, often due to their severe consequences or potential impact on safety. An example is having zero tolerance for safety violations that pose immediate risks to workers. Cross-referencing with "Safety Culture" is pertinent, as zero tolerance aligns with fostering a safety-focused work environment.

CHAPTER 25: CONSTRUCTION CLAIMS AND CHANGE MANAGEMENT

Acceleration: Acceleration in construction claims and change management refers to the intentional speeding up of work to meet project deadlines or recover from delays. For example, if a construction project faces unexpected delays, the project manager might implement acceleration measures to ensure the timely completion of critical tasks. Acceleration is closely tied to "Time Impact Analysis" when assessing the effects of schedule changes on project timelines.

Back Charges: Back charges involve the billing of costs incurred by one party to another party for work or expenses that the first party believes the other should have covered. An example is when a subcontractor fails to meet specified quality standards, resulting in rework costs that the general contractor back charges to the subcontractor. Back charges are interconnected with "Contractual Disputes" as disagreements regarding responsibilities often lead to back charge claims.

Constructive Change: Constructive change occurs when the project owner, through its actions or inactions, imposes changes on the contractor's scope of work or project conditions, even without formal approval. An example is a project owner requesting alterations to the project design without initiating a formal change order. Constructive changes often intersect with "Scope Creep," as they both involve modifications to the original project scope.

Delay Damages: Delay damages, also known as liquidated damages, are financial penalties specified in a construction contract for delays in project completion. For instance, if a construction project is not completed by the agreed-upon deadline, the contractor may be liable to pay a predetermined amount for each day of delay. Delay damages are closely related to "Time Extension Requests" when contractors seek additional time to avoid incurring such penalties.

Eichleay Formula: The Eichleay Formula is a method used to calculate extended overhead costs incurred by a contractor due to a government-caused delay on a construction project. It considers the average daily overhead rate and the number of days of delay. For example, if a government action causes a project delay, the contractor might use the Eichleay Formula to quantify the impact on its overhead costs. This formula is associated with "Government-Induced Delays" in construction claims.

Force Majeure: Force majeure refers to unforeseen and uncontrollable events, such as natural disasters or emergencies, that may excuse a party from fulfilling contract obligations. An example is a construction project facing delays due to a hurricane, allowing the contractor to claim force majeure and seek relief from contract penalties. Force majeure is linked to "Weather Risks" and "Time Extensions" when assessing the impacts of uncontrollable events on project timelines.

Government-Induced Delays: Government-induced delays involve project delays caused by actions or decisions of governmental entities, such as permit delays or regulatory changes. For example, if a construction project experiences holdups due to zoning regulation modifications, the contractor may file a claim for government-induced delays. This concept intersects with "Regulatory Risks" when evaluating delays arising from changes in government regulations.

Home Office Overhead: Home office overhead represents the indirect costs incurred by a contractor to support its overall business operations, such as administrative expenses. When construction projects are delayed or disrupted, contractors may seek compensation for home office overhead costs. Home office overhead is interconnected with the "Eichleay Formula" when calculating extended overhead costs during project delays.

Impact Costs: Impact costs refer to the additional expenses incurred by a contractor as a result of changes or disruptions to the construction project. An example is when unexpected design changes lead to increased labor costs and project delays. Impact costs are closely associated with "Change Order Pricing" when determining the financial implications of modifications to the original project scope.

Job Order Contracting (JOC): Job Order Contracting is a procurement method that allows for the execution of multiple construction projects through a single, competitively bid contract. JOC is beneficial for repetitive work, such as maintenance or repair projects. For instance, a government agency might use JOC to streamline the contracting process for various construction tasks. JOC is connected with "Construction Procurement and Bidding Strategies" when exploring alternative methods for project delivery.

Key Performance Indicators (KPIs) in Change Management: Key Performance Indicators in change management involve metrics used to assess the efficiency and effectiveness of change processes in construction projects. For example, tracking KPIs can help evaluate how well a project team adapts to changes and implements them seamlessly. KPIs in change management are intertwined with "Change Management" itself, as both concepts focus on evaluating and improving the change processes within a construction project.

Liquidated Damages: Liquidated damages, similar to delay damages, are predetermined financial penalties specified in a construction contract for certain breaches, such as project delays. An example is a clause stating that the contractor will pay a specified amount for each day of delay beyond the agreed-upon completion date. Liquidated damages are closely related to "Delay Damages" and are often used interchangeably in contract language.

Mobilization Costs: Mobilization costs refer to the expenses associated with preparing and setting up a construction site, including the transportation of equipment and materials. For instance, when a construction project begins, the contractor incurs mobilization costs to bring resources to the site. Mobilization costs are connected with "Project Startup" when considering the initial phase of a construction project.

Notice of Claim: A Notice of Claim is a formal document submitted by a party in a construction project to notify the other party of a potential claim, dispute, or issue. An example is a contractor sending a Notice of Claim to the project owner to indicate that unforeseen site conditions may result in additional costs. Notice of Claim is associated with "Contractual Disputes" and "Claims Administration" when initiating and managing formal claim processes.

Owner-Directed Changes: Owner-directed changes occur when the project owner instructs the contractor to make modifications to the project scope or specifications. An example is a client requesting changes to the interior finishes after construction has started. Owner-directed changes are interconnected with "Change Orders" and "Constructive Change" when considering alterations initiated by the project owner.

Pricing Methods for Change Orders: Pricing methods for change orders involve the approaches used to determine the cost of changes to the construction project. For example, contractors may use unit pricing, lump-sum pricing, or time and materials pricing for change orders. Pricing methods for change orders are related to "Change Order Pricing" when selecting the most appropriate method for valuing changes in the project scope.

Quantum Meruit: Quantum meruit is a legal term that refers to the fair and reasonable value of services rendered or work performed. In construction claims, it may be invoked when there is no specific contract price agreed upon for additional work. An example is a contractor seeking payment based on the reasonable value of extra work performed beyond the original contract scope. Quantum meruit is connected with "Unforeseen Work" when assessing compensation for work not covered by the initial contract.

Request for Information (RFI) in Change Management: A Request for Information in change management involves seeking clarification or additional details about changes to the construction project. For instance, if the design plans are revised, the contractor may submit an RFI to understand the implications on the project's scope. RFI in change management is associated with "Change Management" and "Claims Administration" when navigating uncertainties related to changes.

Scope Creep: Scope creep refers to the gradual expansion or uncontrolled growth of a project's scope without proper authorization. An example is a construction project initially designed for a certain number of floors but later expanded to accommodate additional floors without formal approval. Scope creep is linked to "Constructive Change" when unauthorized modifications impact the project's original scope.

Time Extension Requests: Time extension requests are formal appeals submitted by contractors to project owners to extend the agreed-upon project completion deadline. For example, if a construction project experiences unforeseen delays, the contractor may submit a time extension request to avoid incurring delay damages. Time extension requests are closely related to "Delay Damages" when seeking additional time to complete the project.

Unforeseen Work: Unforeseen work involves unexpected tasks or conditions encountered during construction that were not part of the original project scope. An example is the discovery of soil contamination requiring remediation during excavation. Unforeseen work is interconnected with "Quantum Meruit" and "Change Orders" when addressing compensation for work not initially anticipated.

Value Engineering in Change Management: Value engineering in change management involves the systematic approach to optimize the value of changes to the construction project. For example, if a design change is proposed, value engineering aims to assess alternative solutions that provide the most value. Value engineering in change management is associated with "Value Engineering" when considering cost-effective alternatives to proposed changes.

Weather-Related Claims: Weather-related claims involve seeking compensation for project delays or disruptions caused by adverse weather conditions. An example is heavy rainfall leading to construction site flooding and work stoppages. Weather-related claims are connected with "Force Majeure" when evaluating the impact of uncontrollable weather events on project timelines.

Xeriscaping (Exterior) in Change Management: Xeriscaping (exterior) in change management involves the modification of landscaping plans to incorporate water-efficient and drought-resistant elements. For instance, if water scarcity becomes an issue during construction, the project team may implement xeriscaping changes. Xeriscaping (exterior) in change management is linked to "Sustainable Construction Practices" when considering environmentally conscious modifications.

Yield Loss Claims: Yield loss claims involve seeking compensation for reduced productivity or efficiency due to changes in the project. An example is the adoption of new safety measures that temporarily slow down construction activities. Yield loss claims are associated with "Impact Costs" when assessing the financial impact of changes on overall project efficiency.

Zero-Cost Change Orders: Zero-cost change orders refer to changes in the project scope that do not result in additional costs for the contractor or the owner. For example, if a change in materials specified in the initial plans does not affect the project budget, it may be considered a zero-cost change order. Zero-cost change orders are connected with "Change Order Pricing" when evaluating the financial implications of changes.

CONCLUSION

As you have delved into the depths of this comprehensive dictionary, you have equipped yourself with an invaluable toolkit of knowledge and understanding, empowering you to navigate the intricate world of construction engineering and management. This extensive resource has served as a beacon of clarity, illuminating the vast array of terms, concepts, and principles that underpin this dynamic and ever-evolving field.

From the fundamental principles of project planning and scheduling to the intricacies of cost control and quality management, this dictionary has provided a roadmap of essential information, guiding you through the labyrinth of construction practices. With each entry, you have gained insights into the language of construction, enabling you to communicate effectively with professionals from diverse backgrounds.

Throughout your journey through this dictionary, you have encountered a wealth of real-world examples, case studies, and industry insights, enriching your understanding of the practical applications of construction engineering principles. These practical applications have not only solidified your knowledge but also equipped you to tackle the challenges and opportunities that arise within the construction industry.

As you embark on your construction engineering and management endeavors, carry with you the knowledge and skills gleaned from this comprehensive dictionary. Embrace the dynamic nature of the industry, staying abreast of emerging trends and advancements in technology and management practices. With dedication and perseverance, you will make significant contributions to the success of construction projects, leaving your mark on the ever-evolving landscape of construction engineering and management.

ABOUT THE AUTHOR

Steven Smith, Ph.d.

Steven Smith is a renowned expert in the field of Construction Management, with a wealth of knowledge and experience spanning both academia and industry. Holding a doctorate in Construction Management, Steven has dedicated his career to advancing the field and contributing to its body of knowledge.

Throughout his academic journey, Steven's passion for understanding the intricacies of construction processes and finding innovative solutions to industry challenges became evident. His doctoral research focused on optimizing project management practices and enhancing productivity in construction projects, leading to a profound understanding of various aspects of construction management and their impact on project success.

BOOKS BY THIS AUTHOR

Dictionary Of Building Materials: A Reference For Building Industry Professionals, Homeowners, And Students

A to Z of Building Materials: The Dictionary You Need

The Dictionary of Building Materials is a comprehensive guide to help you understand the different types of building materials. This essential resource is ideal for building industry professionals, homeowners, students, and anyone else interested in building materials. It covers thousands of building material terms with clear and concise definitions and examples.

Order your copy today and improve your knowledge about the building materials that are used in today's construction industry!

Dictionary Of Building Components: A Reference For Building Industry Professionals, Homeowners, And Students

The Essential Guide to Building Component Terminology

The Building Components Dictionary is an essential resource for anyone who works with buildings, whether as a professional, homeowner, or student. It provides clear and concise definitions of the terms associated with seven essential building components: beams, columns, doors, foundations, floors, roofs, and walls.

Among others, readers will benefit from the following:

Improved communication and collaboration: With a clear understanding of the terminology used to describe building components, you can communicate more effectively with other building professionals and collaborate more effectively on projects.

Informed decision-making: When you understand the terminology used in building construction, you can make more informed decisions about the design, construction, maintenance, and repair of buildings. For example, a homeowner can use the dictionary to make informed decisions about the materials and construction methods used in the renovation of their home.

Broader knowledge of building construction: The Building Components Dictionary is a valuable resource for anyone who wants to learn more about building construction. It provides coverage of a wide range of topics, from the basics to complex technical concepts.

Order your copy today and start mastering the language of building components!

Dictionary Of Contemporary Architecture

Dictionary of Contemporary Architecture: Your Essential Guide to the Language of Today's Architecture

This is a definitive guide to the language of contemporary architecture. The dictionary provides a plethora of key terms, styles, materials, elements, sustainable architecture, and digital architecture. It is an essential reference for architects, designers, students, enthusiasts, or anyone who wants to stay up-to-date on the latest trends and developments in the field.

Contemporary architecture is a diverse and rapidly evolving field, encompassing a wide range of styles, materials, and technologies. To stay informed and communicate effectively with other professionals, it is essential to have a firm grasp of the key terms, concepts, and principles of contemporary architecture.

The Dictionary of Contemporary Architecture is designed to help you do just that. Written in a clear and concise style, this dictionary provides comprehensive and informative definitions of the essential terms in contemporary architecture, from basic concepts like "axis" and "symmetry" to more specialized terms like "bioclimatic design" and "parametricism."

The dictionary is alphabetically organized and easy to use, making it a practical and accessible resource for readers at all levels of knowledge.

The dictionary is intended for a wide range of readers, including: Architects, designers, and other professionals who need a reliable and up-to-date reference guide to the latest trends and developments in contemporary architecture

Students of architecture and related fields who need a comprehensive resource to support their studies

Homeowners, business owners, and other individuals who are interested in building or renovating in a contemporary style
Anyone who wants to learn more about the architecture of our time

Order your copy today and start mastering the language of

today's architecture!

The Dictionary Of Construction Terminologies: A Compendium Of Knowledge For Students, Academics, Practitioners, And House Owners

The Dictionary of Construction Terminologies

Learn the language of construction from one of the most comprehensive dictionaries of construction terminologies available. From architecture and engineering to materials and equipment, this book covers several aspects of construction terminology in clear and concise language. With thousands of entries, the dictionary is an essential tool for anyone who wants to understand the complex world of construction.

The book features:

Comprehensive coverage of construction terminology
Clear and concise definitions, written in easy-to-understand language
Alphabetical organization for quick and easy reference
It is an essential tool for professionals, students, and anyone interested in the construction field.

Order your copy of the book and start mastering the language of construction!

www.ingramcontent.com/pod-product-compliance
Lightning Source LLC
Chambersburg PA
CBHW072201290526
45794CB00004B/1599